New Frontiers in
Science and Technology

New Frontiers in
Science and Technology

Steve Fuller

polity

First published in 2007 by Polity Press
Reprinted 2008

Polity Press
65 Bridge Street
Cambridge CB2 1UR, UK

Polity Press
350 Main Street
Malden, MA 02148, USA

ISBN-13: 978-07456-3693-1
ISBN-13: 978-07456-3694-8 (pb)

A catalogue record for this book is available from the British Library.

Typeset in 10.5 on 12 pt Plantin
by Servis Filmsetting Ltd, Manchester
Printed and bound in Great Britain by Biddles Ltd., King's Lynn, Norfolk

The publisher has used its best endeavours to ensure that the URLs for
external websites referred to in this book are correct and active at the time of
going to press. However, the publisher has no responsibility for the websites
and can make no guarantee that a site will remain live or that the content is or
will remain appropriate.

Every effort has been made to trace all copyright holders, but if any have been
inadvertently overlooked the publishers will be pleased to include any necessary
credits in any subsequent reprint or edition.

For further information on Polity, visit our website: www.polity.co.uk

Contents

Tables and Boxes

Introduction

Science and Technology Studies (STS) is an interdisciplinary field usually defined as the confluence of three fields with distinct intellectual lineages and orientations: *history of science, philosophy of science,* and *sociology of science*. All three have been marginal to their named disciplines because surprisingly few of the original practitioners of history, philosophy, or sociology of science were primarily trained in history, philosophy, or sociology. Rather, they were natural or exact scientists who came to be disenchanted with the social entanglements of their chosen fields of study. In effect, they fell victim to an intellectual bait-and-switch, whereby the reasons they entered science failed to explain science's continued support in the wider society. This point often makes STS appear more critical than many of its practitioners intend it to be. STS researchers are virtually in agreement that people tend to like science for the wrong reasons (i.e. they are too easily taken in by its hype), but relatively few STS researchers would thereby conclude that there are *no* good reasons to like science.

There have been three generations of STS research, and each tells a different story of disenchantment with science. First, the logical positivists, including Karl Popper, were keen to protect the theoretical base of science from the technological devastation that was wrought in its name during World War I. Next, Thomas Kuhn and his contemporaries on both sides of the Atlantic – including Paul Feyerabend, Imre Lakatos, Stephen Toulmin, Derek de Solla Price – tried to do the same vis-à-vis World War II, though, more than the previous generation, they relied on science's past for normative guidance, largely out of a realization of technology's contemporary role in scaling up the scientific enterprise and giving it forward momentum. Finally, the leading lights of the sociology of scientific knowledge – be they aligned with the Edinburgh School (David Bloor, Barry Barnes, Harry Collins, Steve Shapin) or the Paris School (Bruno Latour, Michel Callon, Steve Woolgar) – came of age during the Cold War and the decolonization of the British and French empires. While the Cold War's strong state-centered science policies kept alive what remained of the classical ideal of science governed by a univocal sense of rationality, objectivity and validity, the processes of decolonization presaged science's "postmodern" turn: namely, the devolution of funding and legitimation to reflect local cultural interests. In these diminished circumstances, much like the state itself, science has been forced to explicitly articulate its import in much more specific ways.

There is also a history, a philosophy and a sociology *of technology*, to which a similar response to world events applies. However, it is harder to forge a coherent narrative because many of the people who are now regarded as major contributors to these fields (e.g. Oswald Spengler, Lewis Mumford, Jacques Ellul, Herbert Marcuse, Marshall McLuhan) developed their signature views with an eye more to the general public than specialized academics, even when they were professional academics. In this context, STS has usefully resurrected these earlier inquiries as part of its own emergent focus on science as a "material practice," also known as "technoscience." However, this technological turn tends to underplay the distinctly *ideological* uses of science – that is, what science means and does to and for people, both elite policymakers and mass publics, who are *removed* from its material practices. Most of this book is concerned with redressing the balance on this point.

So far I have recounted the history of STS from an "external" standpoint, but there is also a more familiar "internal" story that STS practitioners tell themselves. This narrative recounts historical and ethnographic studies of "science in action" (Latour 1987) that from, say, 1975 to 1995 gradually falsified normative accounts of science advanced by philosophers, who themselves over this period had become more responsive to empirical (aka naturalistic) arguments. In this respect, the "science wars" of the mid-1990s were perfectly timed, as public questioning about science's disposition in the post-Cold War era coincided with the lowering of philosophical defenses (Fuller 2006a: chs 3–4). However, this internal story, which is now widely shared even by philosophers themselves, ignores the fact that science had been subject to special philosophical attention in the first place more because of what it justifies than what justifies it. On both the technological and the ideological front, science's power as a form of knowledge has rested on its ability to justify practices that might otherwise appear illegitimate. This speaks to the transformative character of science – indeed, as we shall see in these pages, even of the human condition.

I have entitled the book *New **Frontiers** in Science and Technology Studies*, but ultimately there is only one frontier that STS needs to confront. Let me first put it in philosophical shorthand: *what is the normative import of contingency, rather than necessity, as the modality for making sense of science and technology?* Science – and technology regarded as an application of science – has been treated in modern society as the standard-bearer of rationality, objectivity, and validity, the three normative categories most closely associated with "being realistic." This exalted treatment has rested on the idea that science develops according to an internal dynamic that is relatively unaffected by changes in the larger social environment. It still animates most popular science writing, which effectively obscures any awareness of who is already paying for and is most likely to enjoy the benefits (and suffer the costs) of scientific research.

But what happens once science's "internal dynamic" is revealed as nothing but a strategically focused version of the contingencies that affect the rest of social life? Indeed, what if it turns out that the border between science and politics requires constant maintenance by the propagation of authoritative histories and philosophies of science that remain conspicuously silent – indeed, self-censoring – on political matters? This is the frontier that STS needs to negotiate. It helps to explain the field's controversial nature.

There are two basic attitudes one can take toward the contingency of science, once recognized. STS encompasses both of them. The first is that such recognition should not alter our fundamental attitude toward science, since science works on its own terms, a point to which scientists amply attest, even if it could have followed any of several trajectories. It follows that the contingency of science is of "merely philosophical interest," in Wittgenstein's sense of an understanding of world that "leaves it alone." When STS presents itself as an aspiring discipline in search of academic legitimacy, this attitude tends to hold sway. The second attitude interprets such contingency to imply that the course of scientific inquiry could be substantially altered in the future, if only because at various points in the past it could have gone in a variety of directions. In that respect, there is everything to play for by attending to certain features of contemporary science that might promote change in what is perceived to be a desirable direction. When STS presents itself as a political player, perhaps even a social movement, such an attitude toward contingency then prevails.

The contrast in sensibility here is captured in the two senses in which the practices of science might be said to be "conventional": they are either "traditional" or "reversible." Readers not already familiar with my earlier work on *social epistemology* will discover in these pages that my own approach to STS falls squarely under the second category. I see STS as integral to the normative reorientation of science and technology that is taking place in our so-called postmodern times. If "science" still refers to the most authoritative form of knowledge in society, what is *now* the basis for that claim – and what are its consequences for policy, and life more generally? In breach of interdisciplinary tact, I believe that STS can address this question most effectively in terms of a particular chain of intellectual command. (For a precedent from the annals of logical positivism, see the tripartite sequence of tasks for "epistemology," in Reichenbach 1938: 3–16.) This ideal hierarchy would have historians mine the ore of science and technology by recovering its traces from the archives, philosophers smelt it in a usable form as they convert the historians' narratives to durable multipurpose theories and concepts, and sociologists finally market these refined products to policymakers and the general public.

This book is divided into three parts that divide the field of STS into three fundamental problems: *demarcation, democratization* and *transformation*. The first concerns the conceptual space occupied by "science" in our culture. The second concerns the political organization appropriate to science in society. The third concerns the material horizons within which we want science to change our world, including ourselves. Each part successively shifts the burden from the "science" to the "technology" poles of STS.

Part I is concerned with what philosophers call science's *demarcation problem*. The word "demarcation" implies that regardless of how one defines science in conceptual terms, the problem remains of distinguishing it from its concrete rivals and imitators. This problem is itself distinctive to science because of the many traditions that in principle can lay claim to producing authoritative knowledge in society. I divide this problem into science's search for a *revolutionary moment* and a *unifying vision* – on the one hand, how science breaks from rival traditions and, on the other, how science becomes incorporated into a tradition in its own right. These are the respective topics of chapters 1 and 2, which together serve to introduce the reader to the modern history of cross-disciplinary discussions on the nature of science. A theme common to both chapters is the university as the site of a dialectic between science as a social movement and as a disciplinary formation.

One aspect of this discussion carries over into Part II, which is concerned with an STS-inspired discussion of what it means to democratize science in our times. Karl Popper sometimes recast the demarcation problem in political terms as the search for the *open society*, an endlessly self-critical and ever inclusive community of inquiry. But how should such a community be constituted, especially in light of the various historical conceptions of democracy, as well as contemporary demands to incorporate peculiar subjects in the science polity, notably the non-human natural world? Chapter 3 examines the metaphysical bases and political implications of two alternative democratizing strategies: Bruno Latour's *politics of nature* and my own *republic of science*. Although Latour's normative sensibility is *prima facie* more pluralistic than my own, his strategy for constituting a "parliament of things" may foster the dehumanizing consequences associated with the extreme forms of "species egalitarianism" of the animal rights movement. As for my own republic of science, it may be possible only under a limited range of political economies, the implication of which is that the scale and scope of science today is too large to be governable and must therefore be institutionally reduced.

The rest of Part II brings the discussion of democratizing science down to earth. Chapter 4 deals with what I call the *critical deficit of science journalism*, which amounts to the presumption that, politically speaking, the "good guys" do good science, and the "bad guys" do bad science. One of the clearest expressions of this presumption is *The Republican War on Science* (Mooney 2005), a recent best-seller among the disenfranchised liberal elites in US politics. To redress the critical balance, I urge a strong dose of what STS calls "symmetry," namely, unless proven otherwise, the treatment of the ethical character of scientists and the epistemic character of their science as independent of each other. Chapter 5 shifts the focus to the emergence of *research ethics*, especially against the backdrop of the alleged rise in scientific fraud. I argue that the tendency here is to treat problems of science as a social system as if they reflected scientists' personal moral failures. This occludes the genuine problem of *epistemic justice*, which relates to how one arrives at the standard by which scientific conduct is held accountable. In this context, I examine the controversy surrounding the publication of Bjørn Lomborg's *The Sceptical Environmentalist*. This case provides an opportunity for elaborating contrasting regimes of epistemic justice – *inquisitorial* and *accusatorial* – that might be used as the basis for empowering a science court.

Finally, Part III turns to both science's and STS's role as participants in the technoscientific construction of global society – past, present and future. Chapter 6 presents technology as an evolutionarily adaptive feature of virtually all organisms, yet it is only with the advent of science as the West's universal project of Enlightenment that technology starts to be seen as the infrastructure of a genuinely global society. I argue that this project is largely an extended and secularized version of humanity's divine entitlement in the monotheistic religions. This entitlement has been put under severe strain, both metaphysically and politically, with developments in information technology and biotechnology in the second half of the 20th century. I consider how both provide the basis for a revival of the politics of *social engineering*, about which STS has said remarkably little so far. Nevertheless, social engineering will be decisive in how humanity negotiates its position between the divine and the animal, an "essential tension" that STS tends to characterize as the *cyborg moment*, whose Cold War roots are explored.

Part III, and the book as a whole, concludes with a reflexive look at STS's own problematic position in the world today. The field straddles the divide between self-legislating democracy and client-driven consumerism. No doubt STS is a very useful instrument for policy research that produces a frisson by transgressing taken-for-granted distinctions, but perhaps at the cost of becoming a parasite with no intellectual integrity of its own. This chapter is an expanded version of the 2005 annual Nicholas Mullins Memorial Lecture in Science and Technology Studies at Virginia Tech, the university housing the largest STS graduate program in the United States, where I spent four tempestuous but fruitful years in the early 1990s.

Parts of this book have been substantially reworked from a "report on the state of knowledge" I was commissioned to write by UNESCO in 1999 for the introductory volume of its Encyclopaedia of Life Support Systems. I mention this point because its pretext is not unlike the critical survey that Jean-François Lyotard was commissioned to write for the Higher Education Council of Québec in 1979, which resulted in the "postmodern condition" coming to express the temper of our times. However, the spirit of my piece – and this book – is the exact opposite of Lyotard's: I draw on several strands of the history and sociology of philosophy and science to renew the case for knowledge integration in aid of a normatively unified conception of science. My own politics of knowledge reunification has precedent in both the German idealists and the logical positivists. In practical terms, the aim is to promote distinctly knowledge-based institutions, especially the university, as vehicles of democratic social progress of potentially universal scope. STS so far largely stands outside this project. One goal of this book is to bring STS a bit closer to it.

I am indebted to the following people for their input and generosity in providing opportunities for me to develop the ideas and arguments in this book: Babette Babich, Thomas Basbøll, Jim Collier, Bill Keith, Gloria Origgi, Hans Radder, Ravi Rajan, Roland Robertson, Zia Sardar, Jeremy Shearmur, Ida Stamhuis, and Nico Stehr. Also special thanks go to the graduate students and faculty who attended my summer course in 2005 at Virginia Tech, which enabled me to refine and revise my thoughts. Special thanks also to Emma Longstaff and the three anonymous referees at Polity Press, for persistence and patience, in equal measure. This book is dedicated to Dolores Byrnes, a woman whose nature defies description.

Part I

The Demarcation Problem

1

Science's Need for Revolution

A paradoxical consequence of the emergence of a distinct field of inquiry called "Science and Technology Studies" (STS) is it that has helped to undermine the classical justification for just such a field. Originally it was thought that there was something unique about science as a social and intellectual practice that warranted a field, if not exactly STS, then at least relatively autonomous specialities in the history, philosophy and sociology of science. Some described this uniqueness in terms of a set of necessary and/or sufficient conditions that all properly scientific practices share. Others pointed to a mode of succession that characterized an "internal history of science," in terms of which any pretender to the title of science had to demonstrate their legitimate descent (Lakatos 1981). Together the image projected was of a unified conception of science potentially traceable to a canonical origin, aka the *Scientific Revolution*. This classical strategy of justifying science came to be seen in the 20th century as solving the *demarcation problem* – specifically, the problem of demarcating science from non-science, or pseudo-science (cf. Remedios 2003).

The demarcation strategy is familiar from the history of political thought as akin to the genetic basis used to legitimize royal dynasties. However, in the case of science, philosophers sought demarcation criteria that could have been applied across all of history. When something similar has been urged in the sphere of politics, typically under the name of natural law, it has often resulted in calls to overturn the current regime on grounds of illegitimacy. In science, it has resulted in a relatively bloodless coup that now represents the orthodoxy in STS. It consists of a *de facto* acceptance of, on the other hand, a disunified conception of science – or, spun more positively, a recognition of the plurality of "sciences" – and, on the other, the mythical status of a definitive world-historic "Scientific Revolution" (Galison and Stump 1996; Shapin 1996).

A famous 1983 paper by Larry Laudan officially declared the problem's demise (Laudan 1996). There seems to be a broad consensus today among historians, philosophers, and sociologists that science is whatever scientists do – and if they do different things in different fields constituted by recognized scientists, then so be it. Yet, this is precisely the sort of solution that the original statement of the demarcation problem was designed to *prevent*. How can what was so obviously wrong 50 years ago now seem so obviously right? I happen to believe that the demarcation problem is worth reviving today. In particular, there is a need for a "non-providential" account of the nature of science – that is, an account that does not presume that the dominant tendencies in the history of science are *ipso facto* normatively acceptable. STS's rejection of the demarcation problem may be understood as an overreaction that has thrown out the teleological baby with the providential bath water in making sense of the history of science.

This chapter provides an autopsy of the demise of the demarcation problem (cf. Fuller 1988: ch. 7). The first part offers perhaps the most accessible entry point into the problem of demarcation, namely, the historical moment when science came to be formally set apart from other forms of knowledge in society. This is the so-called Scientific Revolution, which allegedly happened in 17th-century Europe. This topic immediately opens up into a consideration of the most influential theorist of scientific revolutions, Thomas Kuhn, especially his impact on STS. In sections 2 and 3, I explore how one might justify demarcation criteria from a historical and political standpoint. Together they constitute the demarcation problem's "social epistemology" (Fuller

1988: esp. ch. 7). Section 2 traces the origins of the demarcation problem to the need to decide between competing definitions of knowledge from a neutral standpoint, modeled on a judgment delivered in a trial. In section 3, I flesh out the politics that inform this backdrop, drawing on Popper's discussion of open and closed societies.

1. The Scientific Revolution: The Very Idea

Although the expression "scientific revolution" is most closely associated with Thomas Kuhn (1970), who embedded the phrase in a general theory of scientific change, it also names a specific time and place – Western Europe of the 17th century – from which descend the modern institutions, methods, theories, and attitudes of science, as epitomized in the achievements of such figures as Galileo, Bacon, Descartes, and, most of all, Newton. Interestingly, the idea of localizing *the* Scientific Revolution dates only to the 1940s, when both the British historian Herbert Butterfield, known for his progressive "Whig interpretation of history," and Kuhn's own mentor in historiographical matters, Alexandre Koyré, an émigré Russo-French philosopher influenced in equal measures by Plato and Hegel, started to speak in these terms (Fuller 2000b: 23).

The use of the same phrase "scientific revolution" in Kuhn's general and Butterfield's and Koyré's more specific senses is only partly justified. The specific coinage was intended to be provocative. It was an anti-Aristotelian and anti-Catholic gesture designed to consign the Renaissance to a pre-modern past that was superseded by the revival of a Platonic theory-driven science (Koyré) and the Protestant Reformation of the Christian conscience (Butterfield). These crucial elements of the modern scientific imagination had been held back by the demands of secular governance and everyday life. Thus, Koyré contrasted two Italians who had been previously seen in much the same light: Galileo's single-minded pursuit of a unified truth marked him as a scientist, whereas Leonardo da Vinci's jack-of-all-trades empiricism did not.

The rhetorical force of the distinction between the likes of Galileo and da Vinci was not lost in the postwar period. In the aftermath of two world wars that implicated science in the manufacture of weapons of mass destruction, the future integrity of science required that it be seen as having historically revolted not only against religion but, perhaps more importantly, technology. Thus, the Scientific Revolution supposedly marks the moment when philosophers came to regard technology as an appropriate means for testing their theories without being seduced by technology's potential as an instrument of domination. In the more metaphysical terms with which both Butterfield and Koyré were comfortable, the Scientific Revolution was about matter coming under the control of spirit, the passions subsumed by reason.

However, the historical identification of the Scientific Revolution causes problems for the periodization of European cultural history that became popular at the end of the 19th century and still prevails, at least in popular treatments. It casts the early modern period as opening with a "Renaissance" that eventuated in an "Enlightenment." The Scientific Revolution supposedly happened at some point between these two epochs – perhaps when they overlapped in the 17th century. Yet, the import of the Scientific Revolution is seriously at odds with the narrative that postulates the Renaissance and the Enlightenment as consecutive stages in history. As represented in Kuhn's *Structure* and elsewhere, the import of the Scientific Revolution is that a group of people, whom we now call "scientists," managed to wrest control of the means of knowledge production from the politicians, religious fanatics, and others who made it impossible to pursue The True independently of The Good and The Just. This autonomization of inquiry epitomizes all the perceived benefits of academic disciplines. They include: (1) secure borders for inquiry that keep larger societal demands at a distance; (2) common standards for incorporating new members and topics, as well as for evaluating their efforts; and (3) discretion over the terms in which the concerns from the larger society are translated into "new" problems.

Yet this "order out of chaos" narrative fails to do justice to the progressive spirit of the figures normally identified with the Renaissance and especially the Enlightenment. These figures – Galileo and Voltaire come most readily to mind – relished whatever immunity from censure they enjoyed but did not generally associate it with the self-restraint, even self-censorship, that is alleged to be a hidden source of power after the Scientific Revolution. Rather, this period (roughly 1400 to 1800) marked the emergence of the *arts of explicitness*, including such wide-ranging pursuits as satire, the quest for a language of pure thought, and indeed, experimental demonstration.

To be sure, the religious wars of the 17th century made Britain sufficiently dangerous to justify the non-sectarian declarations contained in the Charter of the Royal Society (Proctor 1991: ch. 2). However, it is all too easy to project into the past contemporary anxieties about the potential fate of dissident scientists. Indeed, issues of "respect" and "legitimacy" loomed so large in the early modern era because would-be autocrats were often incapable of enforcing their will in the face of resistance. On the one hand, the autocrats lacked the necessary means of surveillance and coercion and, on the other, potential dissenters were not exclusively dependent on a particular autocrat for material support of their work. Together these two conditions ensured that intellectuals could maintain their autonomy by moving between patrons.

The problem of identifying a Scientific Revolution was raised to a problem of global history with another postwar project: the multivolume comparative study of "science and civilization" in China undertaken by the British Marxist embryologist Joseph Needham (Cohen 1994: ch. 6). China was Europe's economic superior until the early 19th century, yet it had never passed through a scientific revolution. Europe's "Industrial Revolution" – a phrase coined in the 1880s, a century after it purportedly began – initiated the systematic development of technology by scientific design. Up to that point, technologies across the world had emerged by means that, for the most part, were innocent of science to such an extent that aspiring innovators had to be accepted into an esoteric craft culture because the relevant knowledge was not seen as the common entitlement of humanity.

In contrast, the idea of science in its modern hegemonic sense presupposes that *all* humans enjoy a privileged cognitive position in nature (that at the moment may not be fully realized), a status associated with the great monotheistic religions descended from Judaism but not those of the East, where humans were seen more as one with the natural world. The idea that humans might transcend – rather than simply adapt to – their natural condition so as to adopt a "god's eye point-of-view," especially one that would enable the "reverse-engineering" of nature, was profoundly alien to the Chinese way of knowing. In this respect, the Scientific Revolution marked a revolt against nature itself, which was seen as not fully formed, an unrealized potential. Francis Bacon's account of experimentation famously expressed this sensibility as forcing nature to reveal her secrets, namely, possibilities that would not be encountered in the normal course of experience.

The idea of humanity giving a divinely inspired reason to nature had become widespread in the West by the late 18th century, especially after Newton's achievement moved philosophers – not least those behind the American and French Revolutions – to envisage society as something designed *ex nihilo* on the basis of a few mutually agreeable principles, what continues today as "social contract theory." In this context, the pre-contractarian "natural" state of humanity appears unruly because its wilder animal tendencies have yet to be subject to a higher intelligence, secularly known as "rationality" (Cohen 1995).

The joining of political and scientific revolutions in this radical sense is due to the Enlightenment *philosophe* most responsible for the rise of social science, the Marquis de Condorcet (Fuller 2006b: ch. 13; cf. Baker 1975). He specifically connected the successful American Revolution and the ongoing French Revolution via the rhetoric of the first self-declared scientific revolutionary, Antoine Lavoisier (Cohen 1985). Lavoisier had recently reorganized chemistry from its traditional alchemical practices into a science founded on the systematic interrelation of atomic elements. However, Lavoisier himself was not an enthusiastic supporter of revolutionary politics, unlike his great English scientific rival, Joseph Priestley, whose radical Unitarian theology forced him into exile in the newly constituted United States, where he was warmly received by the Founding Fathers (Commager 1978: ch. 2). As Priestley celebrated the French Revolution in exile, Lavoisier was guillotined by the revolutionaries at home.

Lavoisier believed that a scientific revolution would stabilize (rather than dynamize, as Priestley thought) the social order. Here he fell back on the classical conception of "revolution," suggested in the Latin etymology, as a restoration of equilibrium after some crime or period of political unrest. Specifically, Lavoisier opposed Priestley's continued support for the practically useful, but logically confused, concept of "phlogiston," the modern remnant of the ancient idea that fire is an ultimate constituent of nature. In this context, Priestley is best seen as an epistemic populist, much like the positivist philosopher-physicist Ernst Mach who, a century later, wanted scientific judgment to be grounded as much as possible in practical experience, as opposed to theoretically inferred entities that only an expert class of scientists might observe (Fuller 2000b: ch. 2).

Kuhn's relevance as a theorist of scientific revolutions emerges at this point – and not only because his own most carefully worked out case of a scientific revolution was the dispute between Priestley and Lavoisier over the nature of oxygen. Kuhn also agreed with Lavoisier that revolutions mainly restored stability to a science – and by implication a society – fraught with long unsolved problems. Kuhn portrays scientists as the final arbiters of when their knowledge has sufficiently matured to be applied in society without destabilizing it. This doubly conservative conception of revolutions reflects Kuhn's definition of science as dominated by only one paradigm at any given moment.

Consequently, despite Kuhn's broad cross-disciplinary appeal, especially among social scientists, Kuhn consistently maintained that only the physical sciences satisfy his strict definition because it is only in these fields (and arguably only until about the 1920s) that scientists are in sufficient control of the research agenda to determine when and how a revolution begins and ends, and its results spread more widely.

Kuhn's conception of scientific revolutions appeared radical in the late 1960s because it was conflated with the then-prevalent Marxist idea of revolution as an irreversible break with the past, something closer in spirit to Condorcet's original conception (Fuller 2000b: ch. 5; Fuller 2003a: ch. 17). This conflation was facilitated by Kuhn's portrayal of scientists in the vanguard vis-à-vis the direction of their own work and its larger societal import. This image was in marked contrast with the perceived captivity of scientists to what C. Wright Mills called the "military-industrial complex."

However, Kuhn's own reluctance to engage with his radical admirers suggests that his model was proposed more in the spirit of nostalgia than criticism and reform. This interpretation is supported by the original Harvard context for the restorative conception of revolution, the so-called Pareto Circle, a reading group named after the Italian political economist Vilfredo Pareto, whose "circulation of elites" model was seen in the middle third of the 20th century as the strongest rival to Marx's theory of proletarian revolution. This group was convened in the 1930s by the biochemist Lawrence Henderson, who taught Harvard's first history of science courses and was instrumental in the appointment of chemistry department head, James Bryant Conant, as university president (Fuller 2000b: ch. 3). In that capacity, Conant hired Kuhn not only as a teacher, which enabled him to develop the more general ideas for which he would become famous, but also as a researcher on the origins of the Chemical Revolution, which eventually gave Kuhn's general thesis about scientific revolutions what empirical credibility it has (Conant 1950).

1.1. Deconstructing the myth of Kuhn as revolutionary

STS's biggest blindspot is its lack of reflexivity (Fuller and Collier 2004: esp. Introduction): unless STS researchers already come to the field with, say, feminist, Marxist or post-colonialist identities, they tend not to reflect on the conditions that maintain their inquiries. To be sure, there was a spell in the 1980s when some prominent British STS researchers devoted considerable attention to a very narrow sense of reflexivity, namely, linguistic self-reference (e.g. Woolgar

1988, Ashmore 1989). In retrospect, it can be seen as a relatively late adoption of Jacques Derrida's "deconstructive" textual criticism, whereby an author is caught in a pragmatic contradiction between the content and the context of her textual utterance. This serves to desta-bilize the meaning of the text, thereby placing the author's authority "under erasure" (cf. Culler 1982). Since most texts self-deconstruct under such intensive scrutiny, STS researchers rarely bother to attend to a more sociologically informed sense of reflexivity, which would interrogate, say, the extent to which STS itself is captive to an air-brushed disciplinary history that it so easily spots in other fields. (An ironist – in the very spirit of the 1980s reflexivists! – might argue that Derridean self-immolation was the only dignified way out of the Thatcherite straitjacket available to British academics.)

Thus, while some prominent STS practitioners (starting with Restivo 1983) have seriously questioned the ideological function served by the popularity of Kuhn's historiography of science, they have failed to alter the general perception that Kuhn turned the history and philosophy of science in the more critical direction from which STS emerged (e.g. Sismondo 2004). This tired tale of Kuhn's ascendancy depicts Kuhn as the one who overthrew the logical posi-tivist hegemony in philosophy of science by demonstrating that science is a historically embedded collective activity that is not reducible to the strictures of mathematical logic and the probability calculus. Yet it is incredibly easy to puncture holes in this myth (my own demolition job is Fuller 2000b). Kuhn never attacked the logical positivists, mainly because he regarded their project as orthogonal, or perhaps even complementary, to his own. Indeed, the positivists were sufficiently pleased with *Structure* to publish it in their own book series. Moreover, at the time of *Structure*'s composition, Kuhn admit-ted to knowledge of only the most general features of positivist doc-trine. Most of what Kuhn learned about logical positivism and its analytic philosophical offspring occurred *after* the publication of *Structure*, once philosophers interpreted the book as relevant to their own ongoing problems and started engaging with its author.

An even more glaring hole in the myth is Kuhn's alleged unique-ness in drawing attention to the historical and social dimensions of science. As a matter of fact, the entire lineage of people we normally call philosophers of science, from Auguste Comte and William Whewell in the 1830s to Otto Neurath and Karl Popper in the 1930s, were preoccupied with the socio-historical dimensions of inquiry. The very use of the word "science" (and its cognates), as opposed to a more generic term like "knowledge," signified the recognition of an activity pursued by many people over a long time, not a solitary individual staring at a fixed object. The open question for these philosophers – many of whom were practicing scientists – was how inquiry should be organized to maximize knowledge production. The various movements that have travelled under the rubric of "Positivism" since the early 19th century have been the main locus for addressing this question, though it has also figured significantly in the major academic philosophical schools – Kantianism, Hegelianism, Pragmatism – and of course positivism's great rival for the political left, Marxism. Moreover, the question has remained the focus of my own project of "social epistemology."

In all of the above cases, it was supposed that the optimal organization of inquiry would expedite social progress more generally. However, as heirs of the 18th-century Enlightenment, these philosophers equally believed that every aspect of the actual history of science should not be treated as normatively desirable – as if every secular error were a sacred virtue in disguise. That would reduce the history to theodicy (i.e. the theological attempt to prove that, no matter how bad things seem, this is "the best of all possible worlds"). Instead, these philosophers held (rightly, in my view) that intellectual maturity comes from recognizing that error is real but reversible. Of course, this sense of "maturity" might entail radical political consequences, especially if one were to argue that scientific – and hence social – progress has been retarded by traditional institutions like the Church or (as Popper's student, Paul Feyerabend, notoriously argued in our own day) the scientific establishment itself.

The key point here – one easily lost when Kuhn is taken to be the source of all philosophical interest in science's socio-historical dimensions – is that one may cultivate a deep and sustained interest in the actual history and sociology of science yet still find much of it wanting, according to standards that are believed (themselves for good socio-historical reasons) to be capable of expediting the progress of both science and society. This was the natural attitude of philosophers of science before Kuhn. It is why the logical positivists, even as they were trying to recast physics in mathematical logic, were also supporting and sometimes even conducting studies into the history and sociology of science.

Indeed, the positivists' "formal" and "informal" projects were interrelated. The need to translate science into a neutral formalism for purposes of systematic evaluation was born of the ideological freight that normal scientific language had come to carry as a result of its secular entanglements with the German military-industrial complex in World

War I. For Germans living with the memory of a humiliating defeat, science represented everything that had been wrong with the war strategy. Thus, as torch-bearers for the Enlightenment, the logical positivists saw their task as preserving the spirit of science from its ideological and technological corruptions, so that scientific hypotheses can continue to be given a fair hearing. In the irrationalist culture of the Weimar Republic, this task may have been futile but it was not based on ignorance of the socio-historical dimensions of science.

However, the image of the logical positivists changed, both drastically and deliberately, once the Nazis forced them – as intellectuals who were leftist, cosmopolitan, and/or Jewish – into exile in the English-speaking world. The ones who migrated to Britain, notably Neurath and Popper, retained their political edge and overarching sense of science as a vehicle of social progress (though Popper's leftism gradually drifted from socialism to liberalism). However, the vast majority who migrated to the United States stuck to the cultivation of scientific formalism, leaving the history and sociology of science for others to pursue as separate fields. This self-restraint may be explained by the positivists' desire for assimilation, which inhibited them from engaging in research likely to lead to a critique of their hosts' socio-epistemic authority. It was in this spirit that the logical positivists, following the example of junior member Carl Hempel, rebranded themselves as "logical empiricists": the old phrase had suggested a continental European conspiracy to use science to launch a social movement, whereas Hempel's neologism evoked more politically correct roots in such genial Brits as Locke and Hume.

As it happened, the logical positivists were not overreacting by going "deep cover," the phrase used for undercover police work when the agents expect no backup if they are caught. Reisch (2005) has shed new light on the paranoid climate of Cold War America, in which many positivists – including their doyen Rudolf Carnap – were investigated by the FBI. Nevertheless, in their American captivity, the logical empiricists brought an unprecedented level of professionalism and technical sophistication to academic philosophy. By 1960, the memories of secular preachers with public missions, like William James and John Dewey, were in American philosophy's mildly embarrassing, pre-scientific past. (This was still the taken-for-granted view when I did my PhD in history and philosophy of science in the early 1980s at the University of Pittsburgh, the last great positivist citadel, where Hempel was one of my teachers.)

Kuhn introduced a domesticated account of the history and sociology of science into the repressive Cold War intellectual environment.

Unlike the original 19th-century visions of science advanced by positivists, idealists, and Marxists, Kuhn's account portrayed the course of organized inquiry as not merely autonomous but, more importantly, *detached* from larger socio-historical developments. In particular, science no longer appeared as the engine of social reform, a potential challenger to the ruling orthodoxy. Rather, science was subject to its own self-regarding cyclical dynamic, the phases of which – "normal science," "anomaly," "crisis," "revolution" – have now passed into the general intellectual culture. To be sure, Kuhn's conclusion that science displays no overall sense of progress caused nightmares for professional philosophers worried about epistemological relativism. But that should not obscure the larger establishment-friendly implication of his account in Cold War America: namely, that (*pace* Lenin) science is not designed to generate a vanguard party capable of posing a radical alternative to the status quo.

As the younger generation of historians, philosophers, and sociologists of science consolidated around the identity of STS in the 1980s, Kuhn came to be seen as the great mythical progenitor – one of the few constants in a field given to rapidly changing fashions. Perhaps the clearest sign of Kuhn's continuing influence is the idea that science and society are taken to be intertwined to such an extent that it is difficult nowadays to speak of the trajectory of science somehow being "ahead" or "behind" that of society at large. Yet, the asynchronous development of entities called "science" and "society" had been essential to the progressive political projects of the modern era, starting with Comte's original Positivism. Usually, as in the cases of Positivism and Marxism, science showed the way for society. However, with the emergence of feminism, multiculturalism, and post-colonialism, science was increasingly portrayed as a vehicle of domination and hence a drag on the emancipation of subaltern groups.

STS stands studiously apart from both of these modernist tendencies by presuming that science is "always already" social in its actual conduct. This posture effectively depoliticizes the study of science by reducing the research site from the scene of conflicting social forces to an open-ended situation, the identity of which is negotiated amongst the parties on site. Here it is worth recalling that STS's main objection to both Kuhn and the dominant school of social science in the Cold War period – structural-functionalism – was that they treated science as an autonomous social sector, not that they portrayed science as fundamentally at home in its social milieu. On that latter point, STS, Kuhn, and structural-functionalism tend to be in agreement.

1.2. Scientific disciplines as social movements in stasis

While the attempt to remake the world in the image of controversial beliefs and desires has always involved risks, visionary inquirers have gladly undertaken those risks, not least because they perceived the world as open to fundamental change – provided that one takes control of the means of the organization and legitimation of inquiry. To be sure, these visionaries recognized that they start as underdogs bearing the burden of proof. Nevertheless, what I would call the "rhetorical porosity" of the world offered hope that the burden may be reversed, as each generation must be persuaded anew that the established way of doing things should (or not) be given yet another lease on life. Of course, the rhetorical task becomes more daunting under certain conditions. These include an escalation in the perceived and real costs of change, and the concentration of resources necessary for its realization. In our "Kuhnified" intellectual world, which is essentially a reification of the Cold War's political horizons, these conditions arguably obtain. Consequently, the sheer persistence of disciplined science, or "paradigms," appears epistemologically luminous in a world that always seems on the brink of cognitive chaos.

Not surprisingly, then, it is common to think that that disciplined science had a rather specific origin that is perhaps even traceable to a singular cultural moment like the founding of the Royal Society – and that its development cannot be, nor could have been, more perspicuous than it has been. Even contemporary philosophy of science, which has almost completely purged its old positivist fixation on the goal of unified science nevertheless refuses to consider that science (or a particular science), had it pursued a different course of inquiry earlier in its history, would have ended up in a better epistemic position than it is in today. It is simply taken for granted that it was better to dump Aristotle for Newton, Newton for Einstein, etc. – and at roughly the times and for the reasons they were dumped. The Popperian philosophers who last questioned these intuitions – Imre Lakatos and Paul Feyerabend – currently reside in intellectual purgatory (Lakatos and Feyerabend 1999). Their awkward status testifies to deeply held assumptions about the metaphysically special character of the history of science *as it has actually occurred*: science normally is as it ought to be. This providentialist position is what passes today for "naturalism" (Fuller 2000b: 263–5).

That the naturalistic mindset first flourished in the period 1620–1770 – that is, roughly between the founding of the Royal Society and the start of the American Revolution – is quite significant.

Alongside the gradual secularization of humanity that followed the Protestant Reformation was a realization that governments of any longevity typically arose from the ashes of war and were maintained by hereditary succession. Succession by election was seen as an opportunity for renewed conflict – witness the intrigues associated with ecclesiastical and academic appointments. At the time, with a few limited exceptions, constitutional conventions remained unproven philosophical fantasies. That autonomous scientific societies managed to survive as well as they did in their self-selecting, self-organizing fashion was thus a considerable political feat in its own right, marvelous to behold and not to be tampered with. It was concluded that the founders of the Royal Society and similar bodies must have therefore hit upon the *via regia* to reality!

Of course, the line of thought I have just described involves a superstitious reading of 150 years of European history, though one that helped to legitimate a "hard-headed" empirical approach to natural history that bore fruit a century later in Charles Darwin's *Origin of Species*. This orientation was especially pervasive in the Scottish Enlightenment, including such icons of modern philosophy and social science as David Hume, Adam Smith, and Adam Ferguson. For them, the human is not in Foucault's Kant-inspired formulation, a "transcendental doublet," half-animal, half-divine – but mere *Homo sapiens*, one clever animal species among many. They held that our capacities for change are inscribed in the variation that our history has already demonstrated and tolerated. Not surprisingly, when human history is seen in this ontologically diminished light, induction emerges as the premier source of knowledge and tradition the source of power. Institutions become entrenched lucky accidents that we radically change at our peril. In the older theological culture that the Scottish Enlightenment was so keen to displace, these contingencies would have been called miracles. However, the Scots did not dispel the miraculous – they merely secularized it. Such a sensibility remains at play in the current fascination with "self-organizing complex adaptive systems," a scientifically updated version of Smith's invisible hand whose standard-bearers include the postmodern philosopher Jean-Francois Lyotard, the neuroscientist Humberto Maturana, the economist Friedrich Hayek, the sociologist Niklas Luhmann, and last but not least, Kuhn.

The flipside of the miraculous nature of paradigmatic science is the rigidity with which it must be maintained – against an impending sense of chaos in the environment. Thus, from reading Kuhn or, one of his key influences, Michael Polanyi, it is easy to get the impression that a scientific discipline is akin to a monastic order in the stringency of its entry criteria, training procedures, evaluative standards, and so on. Nevertheless, until the late 19th century, with the introduction of nationwide textbooks for discipline-based instruction in European universities, a discipline was really little more than a collection of certification boards announcing that a piece of research met the standards that the boards upheld (I include here what is common to doctoral examinations and peer review journals). The exact nature of the training, the source of funding, and the overarching program of inquiry to which the research contributed were largely left to discretion. Of course, some people aspired to stricter criteria, offering arguments for why, say, certain work required certain prior training. The 20th century has been the story of their steady ascent, culminating in the legitimatory accounts presented by Polanyi and Kuhn. But historically speaking, the rigidities associated with Kuhnian normal science have been always difficult to enforce for any great amount of time or space.

Before 1945, before the idea of a miraculous Scientific Revolution, and before the current fixation on Kuhnian paradigms, a looser concept of disciplinarity and its history can be found, one common to, say, the massive studies undertaken by the engineer-turned-historian John Merz (1965) and the neo-Kantian philosopher Ernst Cassirer (1950). Unlike the post-Kuhnian commonplaces of today, according to which disciplines are the natural products of the "functional differentiation" of the cognitive superorganism, disciplines were originally competing world-views designed to explain everything. They flourished as social movements in several countries, where they campaigned against each other to acquire professorships, funding, influence, and so on. "Crucial experiments" and *Methodenstreiten* functioned as symbolic events in the ongoing struggle. Over time, these clashes were institutionally resolved, especially through the creation of academic departments that were entitled to self-reproduction. (The "nebular hypothesis" proposed by Kant and Laplace for the origins of the universe may be the appropriate scientific metaphor here.) In a sufficiently wealthy academic environment, even the losers could console themselves with a department they could call their own. Moreover, the resolutions were themselves subject to significant cross-national differences, such that the losers in one country may turn out victorious in another. As for the apparent "universalization" of particular disciplines – the fact that, say, physics or economics may be taught the same everywhere – that tendency simply tracked the geopolitical interests of the nations whose universities housed the discipline.

I believe that we should return to this older historical sensibility toward disciplinarity, one that diminishes the phenomenon's significance in the ontology of knowledge production. In the older story, disciplines function as little more than the legitimating ideology of the makeshift solutions that define the department structure of particular universities. Taken together across institutions and across nations, the history of disciplinarity constitutes a set of test cases on how to resolve deep differences in cognitive horizons. In effect, today's disciplines were born interdisciplinary, as social movements that aspired to address all manner of phenomena and registers of life, not simply the domain of reality over which they came to exercise custodianship. In this respect, Positivism holds a special place as a metatheory of interdisciplinarity.

Common to the various projects that have traveled under the rubric of "positivism" has been an interest in constructing a medium of epistemic exchange across disciplinary boundaries. Indeed, in the case of the logical positivists, it would not be far-fetched to regard their ill-fated attempts to "unify" science as having taken seriously – much more so than Peter Galison's (1999) bland notion of "trading zone" – that pidgins and creoles may evolve from their origins as trade languages to become the official language of the trading partners. In their original Viennese phase, the logical positivists were keen to invent an interdisciplinary lingua franca from scratch, partly inspired by ongoing efforts in the 1920s to make Esperanto the official language of the League of Nations. A generation later, when the positivist Philipp Frank (1949) sought models for this enterprise in contemporary interdisciplinary social movements, he found two: Thomism and Dialectical Materialism. Both movements, despite their obvious cognitive deficiencies and proneness to dogmatism, earned Frank's respect for keeping alive the ideal of inquiry that roams freely across domains of reality in the service of individual enlightenment and collective empowerment.

Given that Frank published such things as Cold War paranoia was beginning to percolate in the USA, it is perhaps not so surprising that he was suspected of "totalitarian tendencies." Nevertheless, on closer inspection, Frank was simply updating the classical mission of liberal education, which continually forced academics – no matter how specialized their research – to return to the question of what citizens need to know to exercise their liberties most effectively. Indeed, this mission – including its potentially subversive political consequences – could be found in the youthful writings of Wilhelm von Humboldt, the famed first Rector of the University of Berlin. In 1792, at age 25, long before he became the Prussian education minister and icon for its dedicated bureaucracy, Humboldt wrote a Kant-inspired essay, *The*

Limitations on State Action. Here Humboldt entrusted the university with making the state "wither away" (the source of Marx's usage) from a prescriptive agency to a service provider by enabling citizens to legislate for themselves. Humboldt's youthful vision deeply influenced John Stuart Mill, who dedicated *On Liberty* to him. The Mill-Humboldt connection, in turn, inspired Karl Popper to think about epistemological matters in terms of liberal political theory. (My own civic republican approach to social epistemology, developed especially in Fuller (2000a) and Fuller (2002a: ch. 4), carries forward this intellectual lineage.)

Ironically, despite its studied opposition to Kuhn's paradigm-driven image of science, the alternative vision of disciplinarity just sketched happens to be the one that Kuhn's mentor, James Bryant Conant, wrestled with when he created General Education in Science, the program in which Kuhn received his first job, the original testing ground for what became *Structure*'s central theses (Fuller 2000b: ch. 4). The irony here runs especially deep, as it speaks to Kuhn's lack of historical depth about his own social position as an academic: while Conant realized that the university was destined to become the nerve centre of a modern scientific society, academia hardly surfaces as a topic in Kuhn's writings. Once again, paradigms are "sociological" only in the rather shallow sense of being self-organizing collectives. In principle, a paradigm could exist in a corporate laboratory or an academic department. Kuhn lacks not only an account of how scientific knowledge relates to the larger society, but also an account of how one scientific discipline relates to another.

The reason for this peculiarly alienated and modularized conception of organized inquiry is easy to discern. The experimental natural sciences on which Kuhn focuses were not themselves part of the original constitution of universities, nor did they even begin to fit comfortably on campus until the third quarter of the 19th century. In this respect, Kuhn managed to project this uneasy relationship between science and the university onto the entire history of knowledge production. In these postmodern times, when the value of science is much more apparent than the value of universities, Kuhn's vision thus remains a powerful and potentially destructive force on academic life (Fuller 2003a: ch. 12).

1.3. Where is the next Galileo in the postmodern academy?

Given the chequered historical relationships amongst the university, the natural sciences and something that might be called "revolutionary science," are we "right-minded" people mentally prepared to recognize a Galileo in our midst today? In several respects, the postmodern condition of the early 21st century uncannily resembles the intellectual landscape that Galileo faced four centuries earlier, except that the distribution of power has changed. The arguments of Galileo's Catholic inquisitors, especially the Jesuits, sound very postmodern – potentially seeing truth in everything as long as everything is seen in its place. While a quarter millennium had to pass before the Vatican made formal concessions to Galileo, the Jesuits were already teaching in their Chinese missions what Galileo was struggling to defend at home – namely, Copernicus' heliocentric world-system.

Spoken from a position where power is *concentrated*, the Catholic Church's multipurpose epistemic strategy sounds like a revised version of Plato's doctrine of different truths for different classes – only now different cultures are targeted. Thus, the Jesuits would promote Galileo's counterintuitive views of the universe to impress the traditionally implacable Chinese that Europeans possessed valuable knowledge that the Chinese lacked. The Chinese might then become more open-minded toward some of the other cultural baggage carried by the "barbarians," notably Christianity. However, spoken from a position where power is *dispersed*, the idea of double – or perhaps even multiple – truth evinces the pluralism and relativism of the postmodern condition. In both cases, the same thing is denied: that truth is unified – that there is an ultimate arbiter of right and wrong potentially accessible to everyone.

A good indication of how far short the self-advertised radicalism of postmodern "theorists" falls from the threat historically posed by heretics like Galileo is the *de facto* acceptance of pluralism in the academy. The prevalence of mutually incompatible epistemological and ontological assumptions across the disciplines – what, after Kuhn, are too easily called "incommensurable paradigms" – is painfully obvious. College students routinely register this fact whenever they take elective courses outside their major subject, even if they merely stray, say, from sociology to anthropology. But students also learn to discount the cognitive dissonance they experience as symptomatic of their own ignorance rather than a deeper disorder in the organization of academic knowledge itself.

In that respect, today's universities can hardly be accused of being breeding grounds for latter-day Galileos. Some things never seem to change. An unfortunate unintended consequence of the ongoing efforts by universities to retain their corporate integrity has been a tolerance for all sorts of contradictions as long as they are subsumed under a general acquiescence to the academic ethos (which all too often has been defined less in terms of what is included than what is excluded – e.g. politics, religion). This has meant that old and new forms of knowledge coexist – often in the same building – in polite disregard for one another, unless a budgetary crisis forces the disciplines to confront their differences in a mad scramble for scarce resources.

Of course, there have been exceptions to this pick-and-mix, "good fences make good neighbours" policy of *academic syncretism*. The modern era has been marked by several periodic efforts to stage a "Protestant Reformation" in university culture to return it to its roots in the unifying themes of inquiry. Humboldt's founding of the University of Berlin in 1810 on a curriculum underwritten by the philosophy of German idealism is a good case in point. Its demand for research-led teaching and teachable research also animated the early 20th-century efforts of the logical positivists to unify knowledge, but this time on a more democratic basis. Unlike the full-blown idealist schemes of Fichte, Schelling, and Hegel, which had functioned as a nation-building mythology, philosophy would now mediate the reduction of metaphysically inflated knowledge claims to a common currency of logical arguments and empirical observations that would facilitate their validation and consolidation. My own project of social epistemology is very much in the spirit of these unificationist efforts.

Nevertheless, it must be granted that most of the non-academic knowledge-based innovations of the modern era have been explicitly anti-academic. Galileo himself had impeccable academic credentials but found the university's strictures unable to satisfy his desire for intellectual recognition, public notoriety, and financial security. But motives also ran the other way. In Galileo's day, Spain was Europe's most academically saturated society. University attendance had reached a level that would not be matched by Germany and the United States for another 300 years. Not surprisingly, as in our own time, there were more academically qualified people than suitable forms of employment. The spillover constituted the breeding ground for the Golden Age of Spanish literature, most notably that failed job-seeker in the imperial service, Miguel de Cervantes, whose *Don Quixote* invented the ultimate para-academic genre, the novel, whose research component has yet to be properly credited by epistemologists (cf. Collins 1998: 581–2). In addition, more principled academic exclusions had momentous effects. The Royal Society and other early scientific academies were founded because associations with both the dark arts of magicians and the grubbier arts of craftsmen caused experiments to be banned from university grounds. The Industrial Revolution had similar origins, except that its protagonists were from more modest backgrounds and hence more explicitly interested in political reform and economic gain.

Once universities became integral to the reproduction of the nation-state, ambitious people came to be excluded as threats to the legitimacy of either the state or the academy itself. The respective fates of Marx the political agitator and Freud the private practitioner can be largely explained this way. To be sure, Marx and Freud have become academically respectable, but roughly as their threat to the status quo has diminished. Moreover, recent intellectual currents of anti-academic origins, such as feminism and multiculturalism (or post-colonialism), have secured their place in universities largely on the backs of Marx and Freud.

But none of this denies the possibility of breeding Galileos in the current academic climate. If we focus exclusively on the social significance of his research program, and suspend judgment on its actual epistemic merit, sociobiology's founder, the Harvard entomologist, E. O. Wilson, is a candidate for the mantle of Galileo. His 1975 magnum opus, *Sociobiology: The New Synthesis*, attracted the concerted opposition of politically correct biologists and *bien pensant* intellectuals who saw another opportunity to decry the totalitarian uses of science (Segerstrale 2000). Wilson simply said what Darwin himself arguably would have said about the biological bases of social life, had he benefited from another hundred years of genetics and comparative ethology. While that does not make Wilson (or Darwin) right, it does bring to mind Dostoevsky's Grand Inquisitor in *The Brothers Karamazov*, whose burden would have been made infinitely lighter had Jesus not bothered to return to spread the Gospel a second time. In a similar inquisitorial spirit, the rather Orwellian sounding "Sociobiology Study Group of Science for the People," spearheaded by Wilson's Harvard colleagues Richard Lewontin and Stephen Jay Gould, would have wished to let lie the sleeping dogs of Social Darwinism. At the time, attention was focused on real and imagined racialist agendas that might be fueled by Wilson's ideas. However, underlying these concerns were some epistemological views that would have found favor with Galileo's old Jesuit nemesis, Cardinal Robert Bellarmine.

For example, Niles Eldredge, research director at the American Museum of Natural History and friend of Gould, countered Wilson with the idea that evolution occurs at multiple ontological levels, each requiring its own distinct expertise, an account of which must be provided in any comprehensive theory of evolution. These levels are presumed to be equally significant and not simply reducible to "unstable packages of genetic information" (Segerstrale 2000: 137–41). Yet, what ultimately drives Eldredge's overriding respect for "levels" of reality other than strict adherence to the social structure of contemporary biology? Segerstrale dubs Eldredge's stance "Puritan," but Brahmin Hindu may be a more apt religious analogue. My point is that Wilson has been less inclined than his critics to treat disciplinary specialization as metaphysically significant. After all, the most persuasive case that there are indeed multiple levels of biological reality ranging from the biochemical to the superorganic assumes that the distinct methods and theories associated with biology's sub-fields reflect modes of epistemic access appropriate to these levels. From a sociology of knowledge perspective, such an assumption can convert the contingencies of biology's institutionalization into mythology. Nevertheless, this view has come to be the orthodoxy in post-Kuhnian philosophy of science (e.g. Dupré 2003).

However, philosophers of science, because of their own "underlaboring" attitudes to the scientific orthodoxy, may be especially insensitive to Wilson's revolutionary project of reconfiguring the relationship between the biological and social sciences. That two-thirds of today's leading philosophers of biology have allegedly worked with Lewontin clearly does not ease Wilson's rhetorical burden (Callebaut 1993: 450), though Lewontin's adeptness in analytic and formal reasoning, combined with a left-leaning scepticism, must also explain much of his spontaneous attraction to philosophers. The rhetorical bottom line is that Wilson follows the lead of Galileo (who wrote in both the vulgar Italian and the scholarly Latin of his day) in presuming that sociobiological knowledge claims can be discussed openly in both specialist and non-specialist settings (with the same works often addressing both audiences simultaneously), in the course of which errors in theory, fact, and application will emerge and form the basis for improved claims in the future. In contrast, his critics have appeared very suspicious of the "selection environment" for biologically based ideas in contemporary society. They prefer what amounts to a "double truth" doctrine in which sociobiological claims are treated to technical examination in professional settings, while they are prevented wholesale from having any currency in larger policy discussions.

2. The Historical Dimension of the Demarcation Problem

2.1. The identity of science: definition or demarcation?

Introductory logic and critical thinking courses come alive for students when instructors explain and illustrate the "genetic fallacy," which prohibits judgments of a knowledge claim's validity from being based on judgments of its origins. However, to the historian of logic, it is puzzling that this fallacy only comes to be formally identified in 1934, and that scores of philosophers from Aristotle to Hegel and Dewey appear to have flouted it in their own reasoning (Fuller 2000b: 83). Indeed, these philosophers tended to presume that the task of definition requires an appeal to origins. The solution to this puzzle is that the genetic fallacy simply reflects the emergence of multiple, *prima facie* equally plausible, explanations for why people believe as they do, thereby rendering the appeal to origins ambiguous and hence contestable. It therefore becomes necessary to construct a neutral court of appeal that does not presuppose a specific view about the right way of forming beliefs, either in the individual or society at large. Thus, the task of *defining* science metamorphoses into that of *demarcating* true science from its various pretenders. I shall consider in more detail the philosophical moves involved in this transition from definition to demarcation, including the emergence of "epistemology" as a special branch of philosophy in the 19th century.

Questions of identity are typically wrapped up in matters of definition. A definition says what something is, which traditionally has meant how it has come to be what it is. Thus, when Plato and Aristotle defined science, they referred to the state of mind necessary for being scientific. A prime requisite was sufficient leisure to adopt a sustained contemplative attitude toward the world that would enable one to grasp reality in its own terms – as opposed to the terms of one's own interests. To be sure, Plato and Aristotle diverged significantly on how they envisaged this scientific form of knowledge would appear. Plato imagined a uniform geometry, while Aristotle was more comfortable with a patchwork conception. However, both approached the definition of science from a distinctly *genetic* standpoint. Thus, both were exercised by their sophistic contemporaries for making claims to scientific knowledge without having first adopted the appropriate state of mind. The sophists held that the self-interested pursuit of knowledge – say, in aid of winning a lawsuit – is precisely the right attitude to adopt in accordance with the precepts of their own science, which Plato demonized for future generations as "rhetoric." Indeed, seen

through today's eyes, many of the Socratic dialogues can be understood as anticipations of the Turing Test or Searle's Chinese Room,
designed to show sophistic claims to science are ultimately failed
simulations, in which the veneer of scientificity is betrayed by a failure
to grasp the relevant first principles.

This general approach to definition was carried into what the curriculum still calls "modern philosophy." In Descartes' hands, of
course, the relevant first principles reached back to God as the ultimate guarantor of science. The rest of the moderns, not least Kant,
seemed to conflate matters of logic, psychology, and ontology routinely by suggesting that what things are is somehow integrally tied to
how we come to know them. This was the context in which "epistemology" entered English as the name for a branch of philosophy in
1854, courtesy of James Ferrier's *Institutes of Metaphysics: A Theory of
Knowing and Being* (Passmore 1966: 52–3). Ferrier argued that since
what we do not know is by definition knowable, it follows that the
unknowable cannot exist, since everything can be classified as either
known or unknown. Thus, what might from today's analytic philosophical standpoint (especially after Kripke 1980) appear to be a conflation of three different branches of inquiry turned out, in Ferrier's
hands, to provide a coherent foundation for the field of epistemology.

What divides Ferrier from today's philosophical wisdom is that
philosophers now believe that a crucial presupposition of Ferrier's
argument is untenable, namely, that there is a universally agreeable
way of expressing what we know and do not know. The "philosophically correct" view – introduced by Kuhn and Feyerabend and domesticated by Ian Hacking – is that different languages render different
things knowable and unknowable by conjuring them into being
through their saying (Hacking 2002). Thus, the unknowable exists as
the unsayable. But notice that the sense of "unknowable" Ferrier
denied is not the same as neo-Kuhnians now assert. Ferrier was
speaking in absolute terms, whereas the neo-Kuhnians speak rela-
tionally. In other words, for neo-Kuhnians, Ferrier failed to relativize
the unknowable to what is unknowable *if one speaks a certain language
or inquires under a certain paradigm.* But equally, Ferrier could argue
that the neo-Kuhnians hastily presume that natural language or disciplinary jargon provides the right basis for making epistemological
claims. On the contrary, he might continue, a formal logical translation or other mental preparation is required before one can articulate in suitably "universal" terms. Hopefully the reader will appreciate
the link forged here between the classical genetic approach to philosophy and what is usually portrayed as the decidedly anti-genetic

approach of logical positivism. I shall return to this hidden connection below.

While it has become increasingly common, especially since
Foucault (1970), to treat epistemology as symptomatic of a now
fading anthropocentric perspective on philosophy initiated by Kant,
it would be truer to the historical currents to say that "epistemology"
was coined to capture a sensibility that had traditionally dominated
philosophy as a whole but was being challenged in the 19th and, more
explicitly, the 20th centuries. In this respect, the coinage of "epistemology" marked a realization that what had been taken for granted
was now in the process of being lost. Rather than signifying a functional specification of philosophy, the emergence of "epistemology"
pointed to the imminent dissipation of a unified philosophical
purpose. At stake here is an issue in which the interests of metaphysics
and historiography overlap. What to the former appears as an innovation – that is, the coinage of "epistemology" – looks to the latter like
a reminder.

It is worth recalling that according to the Saussurean linguistics
on which Foucault's history of knowledge systems drew, words correspond, in the first instance, to conceptual not empirical objects.
Nevertheless, it is easy to elide the distinction between what are ordinarily called "ideas" and "things" because ideas produce expectations
about the world that then inform the interpretive and physical construction of things. But even granting this elision, we can still wonder
about the *spirit* of the conceptual construction of empirical objects.
Put most crudely: do words introduce new things into the world (in
the spirit of production and description), or do they ensure that old
things are not neglected (in the spirit of conservation and prescrip-
tion)? Thus, Foucault may be right that "man" as a distinct object
of inquiry was not formally identified until Kant's coinage of
"*Anthropologie*" in 1795, yet man may have previously existed. In that
case, Kant may have been the first to recognize the very decline that
Foucault sought to chart nearly 200 years later. Given the decline of
theology as a countervailing force, Kant was trying to hold on to the
uniqueness of humanity before it became completely absorbed into
the emerging naturalistic world-view as just another animal species.

The difficulty of the Kantian task of holding the line on an anthropocentric philosophical standpoint may be charted in terms of the
shift in the dominant meaning of "genetic" from (as it was still in
Descartes) humanity's divine ancestry to our species' emergence from
a common animal nature. In both cases, the appeal to genesis
remained the same – namely, to define our essence by reference to our

origins. (A good source for the shift in the semantic burden of "genetic" from theology to biology is Merz 1965, vol. 2: 280, n. 1; 622–3, n. 1.) This very significant semantic transition is largely obscured in contemporary analytic philosophical discussions of the *a priori* that still claim some pedigree from Plato and Descartes, while trading on more recent naturalized conceptions of innatism. This obscured transition is exemplified in the competing notions of explicit and implicit conceptions of rationality that coexist in the philosophical literature: i.e. one based on self-conscious deliberation (harking to Descartes) and one based on some naturalized sense of economizing (harking to Darwin). Sometimes the two conceptions are fused in a common image, e.g. the application of rational choice theory to animal altruism or Steven Pinker's (2002) embedding of Chomsky's neo-Cartesian generative grammar in a neo-Darwinian evolutionary psychology.

In light of the above history, the 20th-century quest for demarcation criteria of science marked a radical departure from the classical project of defining science. The genetic approach to definition came to be seen as itself a potential source of deception in determining whether something is a science. The prospect of competing theological and biological (and other) origins for science – as for everything else *prima facie* definitive of the human condition – meant that one must specify a neutral position on the basis of which these competing claims could be adjudicated. At this point, the project of definition turned into one of *demarcation*. In particular, the demarcation project was hostile to any default acceptance that a knowledge claim was scientific simply because it had been uttered by someone with scientific credentials or who drew on a presumptively scientific body of knowledge. Although it is common, and certainly not incorrect, to associate the logical positivists who originally proposed demarcation criteria with broadly speaking a "scientific world-view," they kept a studied distance from *naturalism*, which tended to outsource philosophical problems to the natural sciences. Indeed, only the founding logical positivist, Moritz Schlick, called himself a naturalist, while both Carnap and Popper at various points explicitly criticized naturalism (Popper 1959: sec. 4; cf. Rouse 2002: ch. 1).

Interestingly, one of the few people today who shares the positivists' studied neutrality toward naturalism as a philosophy of science is the lawyer Phillip Johnson, the spiritual leader of the scientifically updated version of creationism known as intelligent design (e.g. Johnson 1991). Although it has become common, especially after the US Creationist trials of the 1980s, to cast the demarcation project as especially concerned with discriminating science from religion, religion was the least of the positivists' original concerns in the 1920s and 1930s (Hacohen 2000: ch. 5: cf. La Follette 1983). Rather, their main examples of "pseudoscience" were drawn from Marxism and psychoanalysis, which they saw as attempting to extend legitimate scientific findings into the public domain where they could not be tested properly yet nevertheless could convey the impression that matters were more resolved than they really were. The trappings of science would thus be invoked to restrict the scope of politics by illicitly restricting the sphere of decision-making. Examples of claims likely to have such an effect include "The proletarian revolution is inevitable" and "Your psychic identity has been sealed by age five." Were the original demarcationists alive today, they would probably leave Creationism alone (since it can be – and has been – criticized by ordinary scientific means) and set their sights on the overextension of evolutionary arguments into the human realm, as appear in, say, Pinker (2002). Like the works targeted by the positivists in their original Weimar setting, these too appeal to a sense of "fate" that would turn science into a vehicle for containing rather than empowering people – very much against the spirit of the Enlightenment.

Of course, the positivists did not succeed in designing a "neutral observation language" or "criterion of testability," but the spirit guiding their enterprise is worth sustaining. It aspired to a standard of fairness – or a level playing field – familiar from a judicial proceeding, very much as Francis Bacon had originally conceptualized the scientific method (Franklin 2000: ch. 8). Thus, competing theories would need to be recast in some common medium that avoided any bias that might come from the greater reputation of an incumbent theory, which might benefit not only the incumbent but also the challenger who tries to adapt the incumbent's trappings for its own purposes (Agassi 1983). The distinction between science's context of discovery and justification is the formalist residue of this aspiration. Moreover, a revival of the demarcationist spirit may require that we revisit the modern empiricist doctrine of *tabula rasa*, normally interpreted as meaning a slate that has been never written on. Yet, its true normative force may lie in being a slate whose previous writing has been erased. This would help to explain why the logical positivists were attracted to empiricism at all, since the textbook versions of doctrine equate it with the basest inductivism, which, as Popper most clearly realized, was tantamount to the slavish adherence to tradition. This reinterpretation of *tabula rasa* would also explain the otherwise inexplicable association of British empiricism (especially Hobbes and

Locke) with the version of social contract theory that inspired the American Revolution. Nowadays we would call it "decolonization."

Evidence for the continued attractiveness of the demarcation project is that both undesirable prospects have emerged during the trials over the teaching of Creationism in US public high schools. On the one hand, evolutionists have traded on their scientific authority, taking for granted that any challenger theory must meet *their* discipline-based criteria of epistemic acceptability; on the other hand, Creationists have tried to conceal their own deficiencies by turning the evolutionists' own arguments against them and to their own advantage.

2.2. The autonomy of science: Vienna or Harvard?

A presupposition of the demarcation project that even its original defenders underplayed is that science is not necessarily identical with what the majority of accredited scientists say it is. Such underplaying is indicative of the diminished status of philosophical opinion in the broader academic culture. Although most philosophically generated criteria for demarcating science from non-science have set standards that should make practicing scientists squirm, few philosophers – Popper and his followers are honorable exceptions – have tried to press the point. Instead philosophers have acted as if they themselves needed to cultivate the acceptance of scientists to secure legitimacy.

This problem has been only exacerbated with the rise of an even less secure field, science and technology studies, which tends to track the identity of science through its so-called "material practices" rather than the ends to which these practices might be put. Thus, it is common nowadays to map the ontology of science in terms of "experimental traditions" and "styles of reasoning" that cross-cut what the positivists, Popperians, and even the early Kuhnians would have recognized as either scientific research programs or their technological applications. Inspired by the later Wittgenstein and Alasdair MacIntyre (1984), such traditions and styles focus on "discipline" more as *ritual* than *facility* – "form over function," so to speak. The "autonomy" of science is thus reduced to the judgments of those authorized by the self-governing guilds responsible for accrediting specialists in certain skills and techniques. Absent from this strategy for identifying science is the idea that there might be goals shared by all those who aspire to the title of scientist – regardless of the exact tools of their trade – whose efforts might be judged by those equally interested in the goals but not directly involved in their pursuit. This prospect of an "external" standard of judgment, which effectively casts the demarcationist philosopher as an intelligent member of the lay public, is in danger of being lost today.

In the 20th century, philosophers proposed two historical trajectories by which science could become autonomous from the rest of society. Both are worth bearing in mind in our current predicament. I shall call them, respectively, the *Vienna* and the *Harvard* strategy, which result in a Viennese and a Harvardian sense of autonomy. The names serve as mnemonics for collecting together somewhat diverse figures associated with these locations who held the corresponding positions. According to the Vienna strategy – advanced by Schumpeter, Schlick, and Popper – science emerges once knowledge previously pursued for purely instrumental reasons comes to be pursued as an end in itself. It draws on a mode of genetic explanation common in the Wilhelmine period, whereby an entity that begins with heterogeneous functions develops into several entities with homogeneous functions: i.e. the entity "autonomizes," as in Wilhelm Wundt's explanation for the emergence of Kantian morality or Ernst Troeltsch's explanation for the withering away of the Church. According to the Harvard strategy – represented by Kuhn, Shapere, and the scientific realist Putnam – science emerges once scientists take control of the means of knowledge production and pursue problems they themselves have defined without external interference. In the course of the past century, the Harvard strategy has come to dominate over the Vienna strategy.

Three points of contrast and one commonality are immediately apparent. First, the Vienna strategy dates from before – and the Harvard strategy after – World War II. The former is informed by the expansion of disciplinary specialization in the university system, the latter by the protection that such specialization affords to the research agenda from state and industry determination. Second, Viennese scientific autonomy is tracked by following the function of ideas in the intellectual ecology, whereas Harvardian scientific autonomy is tracked by following the intentions of scientists in self-organizing communities. Third, and related to the second, for the Viennese, science emerges as an unintended consequence of an economically advanced society, whereas for the Harvardians, science emerges as an intentional search for freedom and security from an otherwise repressive and divisive society. Their one commonality is that the founding of the Royal Society of London serves as an iconic event. The Vienna strategy stresses the society's encouragement of discussions of advances in the practical arts and natural observation to drift into second-order reflections, while the Harvard strategy focuses on the society's self-organizing character under the

Crown's legal protection. Significantly, this shared focus quickly moves in opposing directions: Viennese autonomy stresses the counterintuitive consequences that arise from reflecting on the tools of inquiry (e.g. paradoxical properties of the numbers used in ordinary measurements and calculations), whereas its Harvardian counterpart stresses the continuities in enterprise – even amidst changes of focus – that autonomous science permits.

Broadly speaking, the Vienna strategy specifies *economic* – and the Harvard strategy *political* – conditions for the autonomy of science. The former is anchored in the level of overall wealth required to permit a specific group to pursue a new activity, the latter in the collective effort that such a group must make to sustain that activity. One might wish to say that any well-rounded sense of scientific autonomy requires both conditions to be met. Indeed, it is plausible to think that the conditions need to be met sequentially: first economic, then political. However, the two sorts of conditions also pull against each other in interesting ways, largely because they track the identity of science through different entities: the Vienna strategy follows the *ideas*; the Harvard strategy the *people* conveying them. Each leads to its own problems, as discussed below.

The overriding virtue of the Vienna strategy is that it renders the pursuit of science continuous with that of philosophy, at least according to Aristotle at the start of *Metaphysics*, whereby the pursuit of pure inquiry is defined as the prerogative of the leisured. This was not simply meant as an assertion of elitist privilege – though it was that as well. It was also meant to capture a sensibility whereby one would not need to pursue knowledge for some parochially instrumental reason but could be open to what the larger world might have to offer. In such a leisured state of mind, one would not be so quick to classify things in terms that render them to one's advantage but instead would try to appreciate them in their own terms. It meant spending more time and other material resources on inquiry than ordinary laborers could do. The aspect of this justification that was underplayed by Aristotle, at least vis-à-vis his teacher Plato, is that such inquiry is likely to yield insights whose practical significance may not be immediately apparent because they open up realms of being "deeper" than those encountered in everyday life.

"Deeper" is of course a polemical term in this context, since it begs the question of the exact relationship between the labored and the leisured world-views (i.e. the "manifest" and "scientific" image, as Wilfrid Sellars would say). At the very least, a so-called "deeper" understanding of reality purports to classify the world orthogonally to how it ordinarily appears to the senses. Properties of perceptual objects not salient to the ordinary observer are salient to the observer who claims "depth," typically because the deep observer aspires to a more comprehensive vision of the world that leads her to organize a wider range of objects than perception normally affords. A typical case in point is the difference between how a close acquaintance and an evolutionary biologist might regard someone who carries an unexpressed gene for a rare disease. What the former might not even know of that person could turn out for the latter to be her most defining feature as a member of a large population. One might, therefore, say, that the sense of "depth" relevant to the scientific image entails the treatment of people not as fully realized persons but as carriers of theoretically specified properties.

To focus on scientists as persons, we need to turn to the Harvard strategy. In its heyday in the 1960s and 1970s, Dudley Shapere (1984), the leading Harvard graduate among his generation of philosophers of science, declared that science is whatever scientists say it is. This pronouncement was made in a Dewey-cum-Wittgenstein spirit that "being is doing," where saying counts as a "doing." (Hilary Putnam's (1975) slightly later contribution, during his scientific realist phase, was to recover the Marxist praxis-based provenance for this line of thought.) It was also the period when the image of science as a peer-centered activity was perhaps most vivid, not least because of the dominance of the US National Science Foundation as a model for science policy worldwide. (Shapere was the first director of the NSF Program in History and Philosophy of Science.) Nevertheless, taken to its logical conclusion, this strategy implied that the socially recognized self-recognizing community of scientists could legitimately decide that all future hypotheses shall be tested on the football pitch rather than the laboratory. After all, as Kuhn maintained, a crisis in a scientific paradigm is resolved by the long-term statistical drift in scientists' allegiances, which is then retrospectively interpreted as having constituted a "scientific revolution." In that case, could there not be migrations across *methods* just as radical as the migrations across theories that Kuhn originally envisaged? If science is nothing but what scientists do, then methods-driven scientific revolutions would seem to be allowed.

The idea that science might turn into football was presented a quarter-century ago as a "mere" logical possibility with no obvious empirical or normative purchase – except perhaps as a *reductio* of the Harvard strategy. However, over the past three decades, the modal status of this possibility has come to be upgraded. Philosophical intuitions about the univocal character of the scientific method have been shaken by historical and sociological studies that demonstrate the presence of distinct methodological traditions, or "styles of reasoning," which would test – and sometimes have tested – the same theories by radically different means. Moreover, the prospect of science having to make a major shift in its *modus operandi* is now a matter of public discussion, as the excessive cost of traditional laboratory experiments in the physical and biomedical sciences force increasing numbers of scientists to test their hypotheses on computer simulations. This trend has led Horgan (1996) to declare "the end of science" and Wolfram (2002) to herald "a new kind of science." Perhaps Shapere was wrong to suggest that science might turn into football – but it might well turn into a high-grade computer game.

That scientific hypotheses about empirical reality might eventually need to be tested on virtual-reality machines is disturbing for two reasons. First, it suggests that the economic preconditions for the pursuit of free inquiry have not been fully overcome. Indeed, they have re-emerged within the conduct of science itself as a constraint on future progress. Second, the extent of this constraint – the sense of epistemic loss that scientists must absorb as their work is translated from the lab bench to the computer screen – highlights a point to which Kuhn was especially alive in his account of normal science, which in turn made it the object of endless Popperian abuse. The much-vaunted political autonomy of the Harvard strategy amounts to the freedom enjoyed by someone locked inside a house of his own design. The pain of shifting from the lab bench to the computer screen comes from scientists having so closely identified the ends of their inquiry with specific means – or "material practices" – that they cannot easily recognize the same ideas conveyed in a different medium. By the same token, such scientists should equally find it difficult to recognize cognate inquiries from the past that involve significantly different material practices. However, this failure of recognition across historical media should alert one to the prospect that the type of inquiry promoted by the Harvard strategy is more a *religion* (at least in its etymological sense of "ritual") than a science. At the very least, it raises large questions of how to track the ends of science, to which we finally turn.

2.3. The ends of science: providential or corrigible?

Teleology can be conceptualized from two different metaphysical horizons: *providentialism* and *corrigibilism*. The first, providentialism, approximates what economists nowadays call path-dependency. It is the idea that a process is "anchored" by the initially dominant tendency, which establishes an irreversible presumption that guides subsequent development by eliminating alternative trajectories. Providentialism is historically associated with Calvinist theology, whereby humans try to understand their fate in the hands of a creative but inscrutable deity by discerning an overarching trend – be it in their individual lives, society at large, or human history as such. As Max Weber famously observed in his account of the rise of capitalism, Calvinists read the increase in personal wealth as a sign of one's membership in the Elect. This served to reinforce the productivity of such individuals and encouraged others to follow their lead. Induction-based theories of knowledge, especially those presupposing a relative frequency interpretation of probability, are modern philosophical descendants of providentialism. Anti-providentialism is represented by the Calvinists' great 16th-century foes, the Jesuits, who promoted the doctrine of "probabilism," which stressed the capacity of inductive generalizations to be reversed under the right circumstances, if only to demonstrate the undiminished capacity for free will in humans over time, regardless of the decisions taken by previous generations (Franklin 2000: ch. 4).

Originally providentialism helped to explain the epistemic gap between the divine and human mind. Whereas the course of history might look to God like the unfolding of a plan whose outcome is already known, it looks to us like the revelation of evidence on the basis of which we might hope to infer this plan – or at least how we figure in it. However, in the mid-18th century, the epistemic gap between the divine and the human mind began to close, as providentialism became *exclusively* path-dependent. In other words, the prospect that the divine plan's full realization might deviate significantly from what history (properly understood) had already revealed sharply diminished. How and why this came to pass are themselves not fully understood – nor fully appreciated, since we do not seem to live in a period that envisages itself at the verge of unifying all knowledge.

No doubt the impressiveness of Newton's achievement played an important role, especially in the Enlightenment's explicit proselytism on behalf of the mechanical world-view (Passmore 1970: ch. 10). However, the prospect that "the end is near" had an interesting epistemological consequence. By the first half of the 19th century, the philosophies of history proposed by Comte, Hegel, and Marx – none of whom could be called a slavish follower of Newton – shared a much stronger sense of how the past determines the future than of the future so determined by the past. Moreover, this epistemic asymmetry – whereby the causes appeared to be so much better known than the effects – was exacerbated by the followers of these philosophers, who practiced a "politics of fate" that read history as destiny. It was precisely this aspect of the great 19th-century philosophies of history that led Popper to demonize them as "historicist." The point will be developed in the rest of this chapter.

But what exactly is wrong with these secular versions of providentialism? The answer will help us understand the motivation behind the demarcation criteria and the terms of its possible revival today. There are three problems with "neo-providentialism":

1. As time goes on, non-conformity to the dominant tendency appears retrograde and justifiably eliminable, ultimately by force if necessary.
2. Since completion of the human trajectory appears to be in sight, one feels justified in increasing investment in the dominant tendency, even if it results in diminishing returns per unit invested.
3. The ultimate problem is the profound confusion over the exact identity of the endpoint to this trajectory: boredom (à la Nietzsche), suicide (à la Schopenhauer), nihilism (à la Kojève), or utopia (à la Marx)?

The politics of neo-providentialism will be considered in the next section. But note here that extreme providentialism can lead to the sort of behavior characteristic of millenarian or Gnostic movements in their more activist phases: i.e. a relentless pursuit of the end-in-sight that eventuates in the sacrifice of self and others. To remain immunized against this line of thought, one needs to be reminded that our grasp of the divine plan may remain profoundly imperfect, regardless of undeniable improvements in our material circumstances. To their credit, both Comte and Hegel appeared sympathetic to the heretical view that God may need our help to complete Creation because, given the free will originally granted to humans, He himself lacks foreknowledge of the details needed to execute His own plan. Thus, the granting of free will may be understood as a case of "ontological outsourcing," whereby the divine architect contracts with human builders who are permitted discretion on exactly how they meet the blueprint's specifications.

Under the circumstances, it becomes easy to see how teleology's other metaphysical horizon, *corrigibilism*, can look attractive. Instead of providentialism's adherence to the old Leibnizian slogan "whatever is real, is rational," corrigibilism presumes that whatever is rational, is realizable – though rarely realized adequately. Perhaps the most intelligible criterion of realizability is to have been already realized at some time and place – however fleetingly and imperfectly. One then argues that this exemplary episode can be used as the basis for mounting a standard against which to judge all subsequent history, in aid of a more perfect realization of the standard in the future. (The allusion here is to the eschatological function of the persona of Jesus in what, after Jürgen Moltmann, is nowadays called the "theology of hope.") This would certainly explain the positivist and Popperian practice – abhorrent to professional historians – of blithely cherry-picking the history of science for episodes that semi-illustrate their normative schemes, while consigning the rest of history to obloquy, if not sheer neglect.

Behind this practice is the intuition that the ends of science are indeed given by history but not necessarily recognized as such when they first appear. Thus, they are not followed up adequately or, worse, followed up in letter more than spirit, as when Newtonian mechanics was converted into a disciplinary matrix when it should have remained as no more than a Kuhnian exemplar. The implied analogy here is between the monolithic disciplinary matrix and the universal apostolic Church in Christianity (aka Roman Catholicism), on the one hand, and the multiply interpretable exemplar and a multidenominational Christianity as a vocation (aka Protestantism), on the other. It draws on Ernst Mach's self-understanding during his acrimonious debates with Max Planck on the future of physics at the dawn of the 20th century (Fuller 2000b: ch. 2). Philosophy of science then became the standard-bearer of this secularized version of the Protestant Reformation. To be sure, the project has lost favor in our times because philosophers have either come to doubt the existence of such scientific exemplars in history or, more likely, their ability to exemplify them in their own practice.

3. The Political Dimension of the Demarcation Problem

3.1. In search of fallible social vehicles for scientific norms

The demarcation problem may also be addressed in political terms. It consists in the search for a society that can sustain science, such that both can achieve their universalist aspirations – what Popper (1945) called the "open society." How exactly one achieves this ideal – and not settle for some version of the "closed society" – is far from clear. To be sure, there is considerable agreement on the normative structure of the optimal scientific regime. It is epitomized in Robert Merton's (1977) famous four principles: *universalism, disinterestedness, communism, organized skepticism*. They are easily recognized as the sociological correlates of key aspects of philosophical accounts of knowledge. They describe how professional inquirers should deal with each other because knowledge is what they are all dealing with. Such a close fit between social function and epistemic merit has been the source of both the strength and the weakness of Merton's principles. In the past, I have targeted the weakness. Merton's principles are a paradigm case of normativity derived merely from what people say they do rather than their actual conduct (Fuller 1997: 63–7). Just as no sophisticated sociologist would rely solely on the pronouncements of theologians and priests to understand the nature of religion, why should a sociologist of science like Merton take at face value the word of philosophers and the scientists they most admire when it comes to defining the nature of science? Yet, a close look at Merton's footnotes shows that this is exactly what he did.

However, here I wish to draw attention to a silver lining in Merton's sociological cloud. Notwithstanding the methodological deficiencies of his approach, it is striking just how much convergence *does* exist on the normative principles to which lip service is paid, especially given the admitted divergence in scientific behavior across research sites and the larger social projects that the principles have been taken to serve. Recent sociologists of scientific knowledge tend to interpret this fact in one of two ways: either as scientists' naïve recourse to past authority even when it fails to capture their current practice or as their cynical appropriation of such authority as protective coloration for whatever best serves their immediate interests. Naivete or cynicism clearly poses a Hobson's choice to scientists' self-understanding, the rejection of which by the scientific community has spawned the "Science Wars" that have been raging on and off for more than a decade (Fuller 2006a). But if, by contrast, we assume the sincerity of Merton's textual witnesses, it would seem that scientists who otherwise vary in time, space, and discipline have wished to be judged by the same standards. In that case, differences in the conduct of scientists would need to be understood in terms of differences in the local conventions for applying these standards.

The interpretive problem here is not so different from that of judging adherence to the Abrahamic faiths – Judaism, Christianity, and Islam – in lands imposing substantial constraints on religious expression. While such lands are easily envisaged as themselves imposing a religious orthodoxy, it would be more helpful here to think in terms of secular states for which securing borders and commanding loyalty, especially in the face of potential external threats, ultimately take priority over permitting the spread of world-views that derive their legitimacy from a source that is deemed "higher," and hence itself potentially threatening. In that respect, a "scientific community" resembles a "community of faith" in the broadest possible sense (Fuller 2006a: 77). I say "broadest possible sense" so as to distance myself from Polanyi (1957), who interpreted "community of faith" to mean an inward-looking monastic order. To be sure, Polanyi's attitude, so influential on Kuhn (1970), was an effective coping strategy in the West's Cold War research environment, where scientists were permitted considerable freedom as long as they did not question their paymasters' motives or the uses to which their research would be put (Fuller 2006a: 33–4).

However, this quiescent view is only one of two general strategies for realizing the normative structure of science. Again, drawing on the history of Christianity, I have distinguished between "belief by evidence" and "belief by decision" (Fuller 2003a: ch. 10). The former involves a self-certifying personal revelation, the latter a risky judgment that must be regularly tested against the unconverted. A good way to grasp the alternatives is to recall Jesus' best known piece of political advice: "Render unto Caesar what is Caesar's, but render unto God what is God's" (Mark 12:17). Is this a formula for peaceful coexistence or eternal struggle between the secular and sacred realms? If "God" is replaced with some appropriate scientific analogue, say, "Truth," then we can similarly ask whether scientists can responsibly uphold the Mertonian norms while remaining within the rule of law. Or does the rule of law necessarily submit science to a state of captivity against which it must struggle, if not entirely escape? Scientists who feel besieged by "institutional review boards" and "research ethics codes" have a taste for the relevant sense of "captivity."

Interestingly, Merton himself thought he had cut the Gordian knot: he originally proposed his four norms in an argument purporting to show that science can flourish *only* in democratic societies. In his original formulation from the 1940s, Merton was clearly taking a swipe at the Nazis and the Soviets, yet at the time it would have been quite natural to regard these demonized societies as among the most scientifically advanced in the world (Mendelsohn 1989). To his credit, Popper, with whom Merton is often lumped together on this issue, took a more radical approach under the rubric of the "open society," an expression he had adapted from Henri Bergson's speculative sociology of the moral order (Bergson 1935). For Bergson, open societies act on behalf of more than just their own members. They tend to be both egalitarian and expansionist, periodically returning to the ground of their collective being, divesting it of institutional trappings that over time may have led to spurious hierarchies backed by mythic history that then underwrite impediments to free communion.

Bergson had in mind the role of the Protestant Reformation vis-à-vis Roman Catholicism in the history of Christianity. We encountered this sensibility earlier in our discussion of corrigibilist teleology. It had animated the Enlightenment sense of social progress emanating from France and Germany – the two nations where Protestantism's intellectual influence was strongest, albeit (in the case of France) not always successful. To be sure, whereas the Protestant Reformers valorized the personal witnesses and life decisions of true believers, the Enlightenment *philosophes* privileged the sensory experience and experimental trials of rational agents. If Protestantism had encouraged mass Bible reading, the Enlightenment took the revolution in printing one step further by promoting the reading of many books (Wuthnow 1989: ch. 6). Both movements faced the same foe, namely, *epistemic paternalism*, the transfer of epistemic responsibility from oneself to someone authorized to circumscribe permissible belief on one's behalf, notably a priestly elite. This is what Kant derided as the childlike state of "nonage" in which humanity existed prior to the Enlightenment.

Kant believed the Enlightenment marked a turning point in human history. If only he were correct! That the Enlightenment should never be presumed an irreversible achievement is illustrated by the intellectual recidivism of contemporary social epistemologists in analytic philosophy, who have reinvested epistemic paternalism with positive value as "trust in expertise" based on a "division of cognitive labor" that results in a "well-ordered science," presumably for a "well-ordered society" (Goldman 1999; Kitcher 2001). But even had the Enlightenment successfully universalized and secularized the Protestant Reformation, as Kant, Hegel, and Marx wishfully thought, that would not have been sufficient to constitute a Popperian open society. To pose the problem somewhat gnomically in theological terms, the Enlightenment's soteriological vision remained wedded to eschatology. In other words, salvation lay in the realization of fate.

In ancient times, a fatalistic metaphysics was associated with a passive, quiescent attitude toward the world that aimed to minimize suffering in the brief period of our physical constitution. Knowing that the end was near would thus simply trigger a set of behaviors designed to make the transition out of physical existence as painless as possible for all concerned. Here the Atomists and Epicureans of Greece made common cause with the Hindus and Buddhists of India. In the modern era, this mentality is perhaps best represented by David Hume, whose inductively driven skepticism was designed to temper both the scientific and the religious enthusiasms of an increasingly secularized Christian culture. Christianity had raised the classical sense of fate to a second-order concern, that is, an end not merely of particular physical beings but of all such beings. This implied that the world itself had a fate, an end that not only all beings shared but also on which they all converged – an end that is *ultimate*. Moreover, this end was at least in principle knowable because humans were biblically created "in the image and likeness of God" (Passmore 1970).

Several centuries had to pass before the full measure of this idea was appreciated, namely, the period nowadays called the Scientific Revolution, and especially the person of Isaac Newton, whose work still provides the benchmark for human intellectual achievement. Newton inspired the Enlightenment optimism that Hume found ultimately unjustified. The Scientific Revolution retained from primitive Christianity a millenarian sensibility that postulated what has been alternatively called "Judgment Day," the "Second Coming (of Christ)," or simply "Apocalypse" (Webster 1975). This endpoint would launch the rule of the righteous, which, depending on one's theology, may entail domination over those whose lives enjoyed secular success but failed to attend sufficiently to spiritual affairs. In Christian theology, this sensibility is associated with "Gnosticism" (Voegelin 1968). Combined with a proto-scientific mentality, this "end is near" sensibility can have diabolical effects – of the sort that Popper (1957) demonized as "historicism."

The idea of humanity's privileged position to grasp the nature of reality was developed in two increasingly polarized directions in the modern era, depending on whether one conceived of the world (or at least the humanly relevant part) as an open or closed system, respectively. On the one hand, history might be following the trajectory that Newton believed, namely, toward the revelation of the Divine Plan, typically an extrapolation or perfection of ongoing human projects, such as the scientific quest for the maximally comprehensive understanding of reality, which would propel humanity to a higher level of being. On the other hand, history might be moving toward a state of pure entropy, given that in a world where matter is neither created nor destroyed but only conserved, the repeated recycling of physical forms would eventually expend itself in a state of pure randomness, which by the late 19th century had come to be dramatized as the "heat death" of the universe. In that case, humans are unique in the misfortune of possessing knowledge of their fate without the power to alter it. These two species of science-led historicism fostered hypertrophic versions of, respectively, utopianism and irrationalism (Mandelbaum 1971). The more zealous followers of Marx and Nietzsche stand for these equally unpalatable alternatives.

What makes the alternatives unpalatable from the standpoint of Popper's open society is that they invest so much epistemic authority in the latest science that it becomes grounds for suspending the rule of law and its corresponding sense of human dignity. The *epistemological* character of that dignity is worth stressing: the right to be wrong and, reciprocally, the capacity to learn from one's mistakes (Fuller 2000a: ch. 1). Neither is possible in a world where the truth demands acknowledgment from the outset because one presumes it to be largely already known. Under those circumstances, the only decision is whether to fully realize this truth in Marxist revolutionary fashion or to resist it in Nietzschean existentialist defiance. From the Popperian standpoint, both the Marxist and Nietzschean responses at least countenance, if not outright counsel, political violence on epistemic grounds, namely, possession of knowledge of a truth that transcends the understanding of those who continue to uphold established democratic institutions.

The aspect of the Popperian open society that I have been stressing, by contrast, is normally called "fallibilism." It is a profoundly humanistic vision that stands against all forms of Gnosticism, whose cultish appeal has traditionally rested on the prospect that some humans are closer to the divine than others. Fallibilism's epistemological challenge to Gnosticism may be stated as follows: *knowledge can never be so certain as to license a segregation of our fellows into The Elect and The Damned.* In more theologically resonant terms: no one ever comes so close to God as to lose his or her basic humanity. This point has profound implications for how one plots progress toward the open society as an historical ideal.

3.2. Conclusion: the problem of science in open and closed societies

Because *The Open Society and Its Enemies* (Popper 1945) was written and read in the shadow of totalitarianism – first Fascism and then Communism – there has been a strong tendency to treat the difference between open and closed societies as an ideological expression of the distinction between individualist and collectivist social ontologies (e.g. O'Neill 1973). In fact, truer to both Bergson and Popper would be to correlate closed and open societies with realist and antirealist – or "essentialist" and "nominalist," in Popper's terms – attitudes toward *both* the collective and the individual. The scholastic reduction of the difference between closed and open societies to the "methodological collectivism" versus "methodological individualism" debate is effectively biased toward the essentialism favored by partisans of the closed society. Thus, whether one believes that human action is determined by deep-seated structural conditions (perhaps as expressed in the Marxist "logic" of dialectical materialism) or biologically ingrained dispositions (perhaps as expressed in the neoclassical "logic" of constrained optimization), the difference between collective and individual is overshadowed by their agreement that the past overdetermines the future. In contrast, partisans of the open society strive for an attitude toward both collective and individual that remains open to the future – that is, open to novelty, not least in the form of experience that confounds our expectations. In this respect, Popper's policy of piecemeal social engineering can be interpreted as either a collectivist or an individualist position, but in both cases the decision-maker is solely responsible for his or her decision, which is open to revision in light of its consequences.

Popper's early reading of Kierkegaard, translated into German during his youth, probably explains this "decisionism" that reasonably marks his philosophy as *scientific existentialism* (Fuller 2003a: 28; Hacohen 2000: 147–8). The difference between Kierkegaard's radical Protestantism and Popper's scientific existentialism is that the former held that the believer's life was itself a hypothesis about the existence of God, the outcome of which would be revealed only in death, whereas Popper held that humans are distinguished from other animals by our ability to virtualize this process by treating the world as a laboratory and proposing hypotheses whose outcomes can be realized in our lives, thereby allowing time to learn from mistakes (Popper 1972: 122).

The history of closed societies is marked by an increasingly rigid boundary between the collective and the individual, eventuating in the authoritarian-libertarian polarity that came to colonize 20th-century thought. In contrast, the history of open societies is marked by the mutual development of the collective and the individual. The quirkiness of Popper's political philosophy, especially his failure to distinguish between the public and private spheres, may be explained as products of an especially consistent application of the open society mentality (cf. Shearmur 1996: 96).

What I have just suggested as the idealized historical trajectories of the closed and the open society correspond to two narratives of progressive politics in the modern era. The former sees liberty and equality as ideals that pull in opposing directions, while the latter regards neither ideal as fully realizable unless realized together. In more party political terms, the closed society's narrative is about the irreconcilability of liberalism and socialism, while the open society's concerns their reconcilability. In genealogical terms, the former is about escaping Rome at its most papal authoritarian, whereas the latter aims to universalize Athens at its most civic republican. The gradual differentiation of the collective from the individual is an Anglo-French narrative. The mutual development of the two is a Germanic plot familiar to both Marxist and Popperian heirs to the dialectical tradition. More detailed plot summaries follow.

The closed society's progressive narrative: Before the Peace of Westphalia in 1648, the legal fiction obtained that Europe was unified under "Christendom." As this fiction became subject to increasingly divisive interpretations, it lost its grip on the political imagination. But what was to replace it? A geographically bounded version of the Pope's absolute sovereignty, as embodied in a monarch? Or, a secular version of the work normally performed by the papal agents armed with "natural law," namely, conflict resolution when the local parties are incapable of settling their differences? The first option traces the demise of Christendom to the errant ways of Roman Catholicism, but nevertheless insists that a just and orderly society must be closed under a set of (genuinely true) beliefs. This is the source of the communitarian epistemology that informed Calvinist theocracy, Saint-Simonian technocracy, and Soviet bureaucracy. The second option diagnoses Christendom's demise precisely in terms of Catholicism's hubristic insistence on closure under a set of beliefs. Accordingly, piety to God is best shown by tolerance for each sincere believer's mode of spiritual access. Thus, the only allegiance required of citizens is to a set of procedures for resolving conflicts that arise as a by-product of the free pursuit of conscience (later secularized as self-interest). This is the source of the libertarian epistemology that informed Hobbes and Locke, and that underwrites modern liberalism. It follows from this history that socialists and liberals hold irreconcilable views about what is salvageable from the demise of Christendom, and that the modern state is a politically unstable entity ever prone to excesses of totalitarianism or anarchism, as long as both ideologies are in play. Weimar Germany is the obvious case in point.

The open society's progressive narrative: Classical Athens embodied the civic republican ideal of the citizen obliged to speak his mind about the collective interest. Such speech could occur with impunity because a citizen's livelihood – typically a hereditary estate – was itself not open to debate. Thus, citizens had no need to defend their interests openly, and were routinely discredited if they appeared to do so. Indeed, credibility accrued to those who appeared to speak *against* their interests, as a rich person who called for higher taxes. In any case, citizens could be voted down one day and return to contest policy the next. Athens provided, in most respects, a brilliant crucible for forging *res publicae*, "public things" of value beyond the aggregate of self-interested citizens. The one not-so-small problem was that the Athenian ideal was available only to a few whose leisure was purchased on the backs of the majority who remained enslaved. In that case, the political version of squaring the circle is how to preserve a robust sense of Athenian liberty while extending citizenship to an ever larger and more diverse population. The more successful attempts have involved expanding the polity's production of wealth, typically by minimizing the use of human labor, and systematically redistributing its fruits to minimize interpersonal differences in status. However, these attempts have faced three main obstacles: (a) the division of labor that accompanies expansion breeds a *de facto* rule by experts that undermines the sense of equality needed to speak one's mind with impunity; (b) patterns of consumption outpace those of production, resulting in the fiscal crisis of the welfare state; and (c) groups historically removed from the Athenian genesis find their assimilation culturally oppressive. It follows from this history that liberals and socialists basically share the same ends but disagree over the means, the former preferring economic and the latter political innovation. If liberals are reluctant to extend liberties in order to preserve already existent ones, socialists are too willing to reduce liberties in order to extend them to more people.

Let me make explicit the connection between the two progressive narratives and the demarcation problem. Corresponding to the closed society is, so to speak, "closed science," that is, a form of inquiry perceived to have virtually reached consensus on fundamental principles in terms of which society may then be based. The question here is the one that the Enlightenment posed to politics in a wave of post-Newtonian enthusiasm: shall society be organized according to scientific principles or shall individuals be free to turn science to their own advantage? The Constitutions drafted in the wake of the French and American Revolutions, respectively, capture the socialist and liberal options. The relationship between science and society in both cases is unproblematic because nothing that people are likely to do – including scientific innovation itself – would undermine the society's foundations. Science would simply enable each society to flourish according to its own political ideal. Demarcation criteria would help society adapt to change and novelty: is, say, a new finding sufficiently valid to warrant a change in welfare policy, or a new technology sufficiently effective to warrant its commercial availability? As Kuhn might see it: normal science for normal societies.

However, the fly in the ointment of this projected harmony between science and society is the ambiguity of science's "universality": is everyone entitled to knowledge by virtue of their common membership in society or their individual capacity to possess that knowledge? Depending on the answer, some may benefit more than others from science in society, posing problems of justice that socialism and liberalism have solved in their characteristically different ways. I shall return to this line of thought in chapter 3, when I critique Bruno Latour's *The Politics of Nature*.

In contrast, genuinely open societies require an "open science" in which the "universality" of science and society are mutually made. Thus, society's disenfranchised elements can make legitimate claims on the conduct of science, just as science's critical mode of inquiry can raise legitimate problems for the basis of social order. For example, if women or ethnic minorities can legitimately petition the biomedical sciences to alter their research agenda to reflect their interests, then the results of such research can also motivate the institution or removal of discriminations based on gender or ethnicity (cf. Longino 1990). Demarcation criteria function here, very much as Francis Bacon envisaged, for the purpose of adjudicating knowledge claims that inevitably arise as an activity that aspires to universal authority encompasses a larger proportion of the population. Science then serves to sublimate ideological conflicts that, in a non-scientific society, could easily turn violent. Again seen through Kuhnian lenses: revolutionary science for revolutionary societies.

The looming problem here is how to establish a credibly neutral basis for demarcating science in a society whose sense of the universal is always, perhaps radically, in flux. The solution favored by the logical positivists – namely, criteria that require the fewest logical and empirical assumptions – may not work but its rationale should be now clear. From this standpoint, Popper's falsifiability variant of this strategy places the burden squarely on the claimant to knowledge to declare the conditions under which his or her difference might make a difference to everyone.

2

Science's Need for Unity

This chapter is concerned with the vicissitudes of unity as a mark of science. It first surveys the debate over whether science naturally tends toward unification or, rather, requires special philosophical effort (aka methodology) to achieve unification. I then consider the alternative historiographies of science proposed by unificationists and disunificationists, which, broadly speaking, employ a reductionist versus an evolutionary narrative, respectively. Next I examine the systematic misrepresentation (or "misrecognition") of the desire for a unified conception of science by the disunificationists who dominate STS today. Here disunificationists stress an alignment of *what is* and *what ought to be*, aka "the natural" and "the normative," which unificationists typically deny. Unfortunately, this disagreement runs orthogonal to a wider and more familiar epistemological dispute, namely, between realism and constructivism. For example, my own social epistemology is an instance of "constructivist unificationism," though that position rarely registers on the philosophical radar. Consequently, this part concludes with a consideration of the constructivist grounds under which the unity of science might be recovered.

1. The Unity of Science: Natural or Artificial?

There are two broadly different ways of conceptualizing the historical conditions for unifying knowledge. One presupposes that integration is a *natural* development of scientific inquiry, while the other supposes that it is added *artificially* to inquiry's inherently divergent tendencies. These are discussed in this section. Moreover, within each position, a further distinction can be made. Natural unificationists may be either *deductive* or *inductive* in orientation, whereas those who see unification as artificial may regard it as a *perversion* or an *improvement* of the natural course of inquiry. More space is devoted to the artificialist perspective, since that provides the more instructive example for contemporary attempts at unification.

1.1. Unity as natural: deductive and inductive versions

When unification is seen as a natural development of knowledge, it is usually in one of two ways, both of which were developed in Western Europe in the second quarter of the 19th century. The deductive version is associated with Auguste Comte's understanding of positivism, whereas the inductive version is associated with William Whewell's vision of science as what he called "consilience." According to the deductive orientation, unification occurs by the most progressive discipline serving as the methodological and theoretical template for expediting the development of more backward disciplines. This strategy, recognizable as a version of what I call in the next chapter the "instantiationist" approach to reality, has historically relied on Newtonian mechanics as the template. In contrast, the inductivist imagines that unification is an emergent feature of several strands of inquiry flowing into a common trajectory, much as tributaries flow into a major river (I shall return to this image in section 4). Here too Newtonian mechanics serves as the paradigm case, but less as an exportable model than as the synthesis of several prior and often countervailing tendencies, as in Newton's unification of Baconian empiricism and Cartesian rationalism. The neo-Darwinian synthesis of the 1940s may be similarly portrayed as having brought together the natural history and experimental genetics traditions in biology, which represent, respectively, conservationist and interventionist attitudes toward the physical environment.

The different roles that Newton's work plays in the deductive and inductive orientations reflect an underlying difference in the institutional location of the people who laid claim to unifying the sciences. From today's standpoint, deductivists are best seen as self-styled "knowledge managers" who took it upon themselves to instruct and encourage inquirers in the backward disciplines to approximate Newtonian standards. The leading 19th-century positivists – Comte, John Stuart Mill, and Herbert Spencer – were freelance writers more warmly received by chemists, biologists, and social scientists than by the professional physicists whose practices they wanted their readers to imitate. In contrast, inductivists have tended to be ensconced academics who used the classroom as the medium for bringing together disparate strands of thought, much as Newton himself did at Cambridge. Whereas the deductivists pitched their arguments for integration to mature knowledge producers, the inductivists held that unification was always a project best left for the *next* generation of disciplinary practitioners, since the value of unification would lie in the fertile ground it laid for future research, as opposed to the legitimation of current research.

It should be noted that the natural approach recognizes that the unification of the sciences may be artificially *blocked* by various means, including disciplinary parochialism and ideological opposition, as well as the fact that genuinely new discoveries and inventions are difficult to assimilate into the collective body of human knowledge. Despite being subjected to severe criticism for his own sociobiological commitments, E. O. Wilson (1998) has been exceptionally sensitive to all these blockages. He suggests that the very development of the social sciences as autonomous fields of inquiry has constituted one such long-term blockage. Before Emile Durkheim's (1964) 1895 declaration that social facts are "*sui generis,*" the social sciences had routinely sought a systematic understanding of human nature, in which biological considerations informed at least the foundational principles. This train of thought connected Aristotle, Montesquieu, and Adam Smith with Marx and Wilhelm Dilthey, both of whom embraced Darwin's theory of evolution as underwriting the unity of *Homo sapiens* as, respectively, producers and interpreters of their life conditions.

To be sure, the autonomy of the social from the biological sciences has been also spurred by the rediscovery and continuation of Gregor Mendel's original work in genetic inheritance, against which social scientists often defined their research in the 20th century, via the "nature vs nurture" controversy. Before the rediscovery of Mendel's work in 1900, the germ plasm was seen as susceptible to environmental (and hence, social) influence. But with the rise of an autonomous science of genetics, biological and social factors have been increasingly seen as trading off against each other, perhaps via some "epigenetic" process whereby the environment triggers preordained patterns of behavior. Consequently, sociobiologists hold that social scientists provide much data, but little theory, that is of use to the overall synthesis of human knowledge because social scientists willfully ignore the biological bases of human life. For their part, social scientists respond that the institutional and political cost of the sociobiological synthesis is much too high, namely, an elimination of the modes of intersubjectivity that precisely distinguish humans from other animals.

1.2. Unity as artificial: positive and negative accounts

To say that the unification of the sciences is "artificial" is to admit that inquiry naturally tends toward dispersion and fragmentation, unless specific measures are taken to alter that tendency. The historical intuition behind this vision is that the special sciences have successively spun off from their original basis in philosophy by "disciplining" inquiries that had been previously subject to unresolvable metaphysical or ideological disputes. On this view, the most advanced sciences are the ones that left philosophy first, whereas the least advanced sciences retain a strong philosophical residue. This distinction is epitomized in Kuhn's idea of "paradigm," a model for inquiry in the physical sciences that is generally lacking in the social sciences, in which the most popular theoretical frameworks (Marxism, Freudianism, behaviorism, cognitivism, sociologism – not to mention capitalism) also happen to be the ones over which there is the most disagreement.

Like the natural approach to unity, the artificial one is also largely a product of a 19th-century vision of European intellectual history, but one bred in the fourth quarter of the century, once academic philosophy began to formally acknowledge the devolution of its inquiries to specialized departments. Artificialists who treat unification as an improvement on the natural course of inquiry have generally envisaged the university as an agency for consolidating the power of the nation-state; whereas those who treat unification as a perversion of inquiry have envisaged the university's agency in more politically modest terms that have often afforded the easy appropriation of academic knowledge for non-academic ends. In the case of "pro-unificationists," the open question is whether an intellectually consolidated nation-state encourages more inclusive or more discriminatory policies toward its own residents. In the case of "anti-unificationists," the open question is whether the appropriators represent the entire public or merely private interests.

1.2.1. Artifice as positive: from sublation to reduction

The idea that the unification of the sciences requires deliberate effort in the face of natural dispersion is traceable to the role that the great synthetic idealist philosophies of the early 19th century played in reinventing the university as the founding institution of German national unity. The figures behind these philosophies – Fichte, Schelling, and Hegel – were all prominent civil servants. Common to their quest for synthesis was a belief that humanity had suffered a "fall" comparable to the expulsion of Adam and Eve from the Garden of Eden. The Tower of Babel – though referring to a later episode in biblical history – became the symbol of this decline. The idea of "pristine wisdom" (*prisca sapientia*) available to the ancients but lost to the moderns had motivated the Italian Renaissance's recovery of the founding languages of Indo-European culture, which culminated with the rise of philology as the intellectual-cum-ideological basis for Europe's primordial Aryan identity in the late 18th century.

However, even the most zealous unifiers of the sciences have rarely believed in a return to an original holistic state of inquiry. Indeed, unification always comes at a cost. For example, the principal Hegelian mechanism of integration, *Aufhebung* – translated in English as "sublation" or "sublimation" – implies that disparate bodies of knowledge must lose some of their distinctiveness to become absorbed into a greater synthesis. By the end of the 19th century, *Aufhebung* itself had come to be sublated into a more generalized concept of *reduction*. What was usually claimed to be lost in "reduced" sciences were the marks of their spatio-temporal origins, which had restricted the range of both eligible contributors and validators of knowledge claims. For this reason, reductionism has tended to be championed by anti-establishment, often younger, inquirers with relatively little investment in the current epistemic orthodoxy (Mendelsohn 1974). This pattern fits not only the major idealist and materialist movements of the 19th century but also the most famous pro-unity movement of the 20th century, logical positivism.

Perhaps the most interesting recent attempt to justify the reductionist approach to knowledge integration is due to Alexander Rosenberg (1994), who argues that a combination of human cognitive limitations and instrumental interests renders reductionism an inevitable feature of any form of knowledge that aspires to understand stable patterns of phenomena. However, from the standpoint of the history of science, what ultimately tends to be "reduced" is the need for an exact physical specification of the phenomena in question. Thus, the fundamental categories of the non-physical sciences in the strict sense – that is, the biological and social sciences – are typically defined by the *functions* they serve in a larger system, not by properties intrinsically possessed by the individuals falling under these categories. For example, a species in biology is defined in terms of organisms that can conjoin to produce fertile offspring, regardless of the organisms' exact physical constitution. This means that in principle *Homo sapiens* could be perpetuated by individuals with very different physical make-ups, if they could perform the requisite procreative acts. Of course, naturally occurring tendencies in genetic variation and considerable inter- and intra-species genetic overlap make this unlikely – that is, just as long as biotechnology can do no more than enhance or diminish tendencies already present in the gene pool. But once that technical barrier is surmounted, the biological definition of species would seem to allow certain high-grade silicon-based androids to pass as humans.

It follows that the reductionist imperative ultimately renders the distinction between idealism and materialism itself immaterial. Inquirers' interests in adopting a particular theoretical system ultimately override the difficulties in specifying all the physical parameters needed to identify entities in that system. In short, physicalism turns into functionalism as a matter of convenience. More precisely, *micro-reduction* eventually yields to *macro-reduction* as the preferred unification strategy. For example, in the late 19th century, the history of biology was commonly portrayed as a battle between "mechanists" and "vitalists" (Cassirer 1950: chs 10–11). The main theatre of war was the ontogenesis of the individual organism. The mechanists fought best from the controlled space of the laboratory, where they could demonstrate the sequence of stages by which an organism is constructed, while the vitalists preferred the open field, where the developmental pattern of organisms seemed capable of adapting to changing environmental conditions. By the early 20th century, the battle had been projected on a broader ideological canvass between "mechanistic" geneticists and "vitalistic" evolutionists, which was formally resolved in 1937 with

Theodosius Dobzhansky's publication of *Genetics and the Origin of Species*, which canonized the neo-Darwinian synthesis (Smocovitis 1996). Dobzhansky's great rhetorical achievement was to cast the experimental geneticist as simulating natural selection in the construction of laboratory conditions under which mutations would be induced (Ceccarelli 2001: chs 2–3). Generally speaking, the initial empirical successes of the mechanists yielded over time to the vitalists' more comprehensive vision. This compromise is signaled by the pervasiveness of the language of *design* in contemporary biology, which includes such related quasi-teleologcial notions as "adaptation," "fitness," and, of course, "function" itself.

This, in turn carries profound normative implications. Consider, by way of illustration, the history of systemic attempts to understand the nature of (1) *mind* and (2) *society*.

1. Systemic attempts to understand the mind began as a battle between those who wanted to reduce mind to matter and those who held that mind transcends matter. Over the last 250 years, the former research program's achievements have rendered the latter program less plausible. But ultimately, difficulties in specifying all of the mind's physical parameters have led inquirers to think in more functionalist, sometimes even reified, terms (e.g. Pylyshyn 1984 on "cognizers" as a natural kind). Increasingly, then, mental life is defined "formally" so as to be indifferent to whether it transpires in carbon-based brains, silicon-based computers, or some combination of the two (i.e. "cyborgs": Haraway 1990).
2. Systemic studies of society began as a dispute between those who would reduce society to the physical circumstances of its members and those who held that society's existence transcends those circumstances. The "naturalistic" tradition from Aristotle and Ibn-Khaldun through Hobbes and Spinoza to Marx and Darwin has cast significant doubts on the idea of societal transcendence. Yet, this tradition's own shortcomings have encouraged the notorious philosopher of "animal liberation," Peter Singer (1975), to redraw the boundaries of the social order so that some (sentient) non-humans are included, while some (disabled, deranged) humans are excluded. Functionality here is defined in terms of the capacity for a flourishing life, regardless of species.

Just as a brain is no longer seen as either necessary or sufficient to constitute a mind, we now seem to be on the verge of accepting that the presence of humans is neither necessary nor sufficient to the constitution of society. Whatever one makes of the normative significance of these developments, one thing is clear: physical requirements eventually yield to functional ones for purposes of constituting a system. The person in our times who did the most to use this point to realize Comte's dream of rendering social science the queen of the sciences was the polymath, Herbert Simon (1977). Simon spent the second half of the 20th century translating the concept of "bounded rationality" from the decision-making methods of harried bureaucrats to the discovery procedures of programmable computers, two contexts united by a need to solve specific problems within limited resources (time, money, personnel, processing capacity). Although he failed to reposition the social sciences as the "sciences of the artificial," Simon was rewarded with a Nobel Prize in Economics in 1978 for having demonstrated that people seek satisfactory, as opposed to optimal, solutions to pressing problems because of their concern with solving similar or related problems in the future, realizing that time is a relevant dimension for measuring resource scarcity (my own PhD was concerned with the implications of this point for scientific and legal decision-making: Fuller 1985).

1.2.2. Artifice as negative: from Kant to Kuhn

Those who regard unification as a perversion of the natural course of inquiry generally object to unification's historical tendency to reduce differences in bodies of knowledge to differences in how one accesses and/or expresses a common reality. This is another way of saying that unificationists tend to collapse ontology into epistemology. Three contrasts capture what has been at stake in the opposition to unificationism. First, whereas unificationists have aspired to one ultimate court of epistemic appeal – say, by reducing all knowledge claims to statements about sensory experience or physical evidence – disunificationists have celebrated the proliferation of local knowledges. Second, whereas unificationists have been keen to remove barriers to epistemic access associated with jargon and related mystifications, disunificationists have aimed to protect spaces for different epistemic communities to flourish autonomously. And third, whereas unificationists have sought to provide a common direction to disparate bodies of knowledge, disunificationists have demythologized historical narratives that promote just this sense of teleology, so as to enable different fields to follow the path of inquiry wherever it leads.

When disunificationism was first articulated as neo-Kantianism in Germany in the third quarter of the 19th century, the idea of a unified sense of reality was criticized on the grounds that it was, in Kant's original term, "noumenal," which is to say, knowable only through its appearances and not in itself (Schnaedelbach 1984). For the neo-Kantians, these various appearances were systematically compre-hended by the academic disciplines that had begun to spin off from philosophy and establish themselves as university departments. The spirit of this argument may be captured by the slogan: *scientific episte-mology recapitulates academic bureaucracy.*

Disunificationists nowadays take a somewhat different tack. They normally suppose that reality is indeed knowable, but it happens to be diverse in nature. Taking the long historical view, they forgo Kant in favor of Leibniz and ultimately Aristotle. However, its recent incar-nation is due to Stanford University in the late 1980s, when the phi-losophy department housed Ian Hacking, Nancy Cartwright, Peter Galison, and John Dupré (the legacy is presented in Galison and Stump 1996). Following Kuhn (1970), they regard the history of science as the natural history of human knowledge. Thus, the fact that scientific paradigms operate on self-contained domains of objects that are accessed by unique sets of concepts and techniques implies the existence of multiple incommensurable realities. A slogan that cap-tures the spirit of this perspective is: *scientific ontology recapitulates laboratory technology.* Another important difference between the neo-Kantian and neo-Kuhnian versions of disunificationism is that whereas the neo-Kantians tended to envisage disciplinary specializa-tion as akin to the "functional differentiation" of organs in the matur-ing embryo, the neo-Kuhnians appeal to a biological metaphor that renders knowledge production irreversible but completely purpose-less, namely, neo-Darwinian speciation, which is, in Kuhn's terms, a "progress *from*" that is not a "progress *to*."

Neo-Kantianism arose in response to philosophy's loss of authority as the interface between the university and its state sponsors (Collins 1998: ch. 13). This is usually traced to the politically divisive conse-quences of the ideological uses made of Hegel's unified vision of reality by his followers in the 1840s ranging from the critical theologian David Friedrich Strauss to the political economist Karl Marx. Thus, instead of discussing the larger societal ends of knowledge, as Hegel had seemed to encourage, the neo-Kantians confined the normative purview of academics to the "peer review" of their own disciplines, the results of which then could be appropriated as the state saw fit (Proctor 1991: Part 2). Even philosophers were reduced to "underlaborers" for the special sciences, who spent their time disentangling disciplinary foundations, streamlining disciplinary histories, and adjudicating disciplinary boundary disputes – again, all with the aim of making

academic knowledge more accessible to non-academic audiences
(Fuller 2000b: ch. 6). Not surprisingly, neo-Kantianism is the source
of the ongoing debates over the sorts of knowledge that the natural sciences, social sciences, and humanities make possible. Wilhelm Dilthey,
Max Weber, and Ernst Cassirer are figures from this movement whose
work in this vein continues to have currency (Habermas 1971).

After Germany, the world's undisputed scientific leader, was
subject to an unequivocal humiliation in World War I, an irrationalist
backlash followed, which was crystallized in Oswald Spengler's
(1991) popular *The Decline of the West*. The neo-Kantians had no
effective response to this backlash, since they had divided the forces
of "Reason" and "Truth" into specialized "reasons" and "truths" that
together evaded Spengler's looming question, namely, what precisely
gives world-historic meaning and direction to the pursuit of knowledge. They became captive to their relativism. Thus, the Weimar
Republic anticipated the postmodern turn with its pluralistic embrace
of such "alternative" sciences as psychoanalysis, homoeopathy, and
even astrology. In contrast, the followers of logical positivism, critical
theory, and existential phenomenology tried to tackle the normative
questions posed by Spengler in their own distinctive ways, leading
thinkers as different as Karl Popper, Theodor Adorno, and Martin
Heidegger to trawl through the history of philosophy to find lost
insights into the unity of inquiry, fully realizing that they might appear
"reductive" to those who treated the current array of academic disciplines as normatively acceptable (Fuller 2003a: chs 14–16). In effect,
they stripped modern science of its diversity and technical virtuosity
and brought it back to basics. Thus, Popper and Adorno returned the
sciences to their original unity with Socratic dialectic in the spirit of
critical inquiry, while Heidegger sought an ultimate sense of "Being"
that is presupposed by the various "beings" studied by the sciences.

Perhaps unsurprisingly, recent disunificationists have cast this
return to unity as a big mistake, exemplified by the dogmatic con-
frontations between "analytic" and "continental" schools of philosophy that marked the second half of the 20th century. However, the
most plausible alternative historical trajectory recommended by the
disunificationists would involve following the lead of the last neo-
Kantian, Ernst Cassirer (Friedman 2000). Unfortunately, not only
did Heidegger handily usurp Cassirer's Kantian inheritance at their
famous confrontation in Davos, Switzerland, in 1929, but more deci-
sively Moritz Schlick – the convenor of the Vienna Circle – questioned
the wisdom of a philosophical position like Cassirer's that endlessly
adjusted itself simply to fit the latest scientific fashion: is justice done

to either Kant or Einstein, if Einstein's counterintuitive physical theories are assimilated to Kant's down-to-earth Euclidean epistemology? In terms of Weimar politics, neo-Kantianism was subject to a deadly pincer attack from philosophical radicals on, so to speak, the "right and "left": the existentialists who wanted to bring philosophy back to theology, and the positivists who wanted it to catch up with science. While Heidegger wondered where is the place of authentic being in an epistemic regime that enables anything to be studied by any means, Schlick wondered at what point does philosophy's relationship to science become so attenuated as to be arbitrary and hence merely ideological "nonsense."

It may not be long before neo-Kuhnian disunification is subject to a similar backlash. Creationism (or intelligent design), ecologism, sociobiology (or evolutionary psychology), feminism, post-colonialism, and computationalism (or virtualization, complexity theory) are very different contemporary science-oriented socio-epistemic movements that have, with varying degrees of success, criticized the disciplinary structure of universities for evading matters of "ultimate concern" that come from explicit recognition that a variety of people and things inhabit a common reality. Each movement is poised to become a new source of unificationism. Table 2.1 summarizes the differences in the unified and disunified images of science, as a prelude to considering the systematic misrecognition that today's disunifiers perpetrate against a renascent unificationism.

Table 2.1. Unified vs disunified images of science

Image of Science	Unified	Disunified
Role of philosophy	Prescribe ideal norms of "science"	Evaluate within existing scientific norms
Fact/value	Clearly distinguished	Blurred or deconstructed
The import of "naturalism"	Critical of science's metaphysical pretensions	Clarifies the nature of scientific authority
The source of epistemic authority	Concentrate sovereignty in "science"	Distribute sovereignty to many local knowledges
Attitude toward disciplinary boundaries	*Anti:* They provide barriers to universal epistemic access	*Pro:* They provide spaces for epistemic communities to flourish
The role of philosophy of science	Provide direction to disparate bodies of knowledge	Free bodies of knowledge from a preordained trajectory

2. The Misrecognition of Unity in Recent History and Philosophy of Science

With the end of the Cold War, the image of unified science has lost its previously compelling nature among both philosophers and scientists (Fuller 2000b; Mirowski 2002; Reisch 2005). However, there are very good reasons to defend a specifically *constructivist* version of unificationism and reject the fashionable image of science depicted by the realist disunificationists. Specifically, realist disunificationists align "the natural" and "the normative" so as to allow scientists to avoid their social responsibilities, whereas the constructivists embrace science's social responsibilities, as per the "artificialist" perspective discussed in section 1.2 above. I shall explain and defend this point more fully in section 3. But first, in this section, I present the image of the unity of science program in the eyes of its disunificationist historians, highlighting the shortcomings the disunificationists reveal about their own philosophical position.

2.1. The gospel according to the disunificationists

History is full of cases in which something is defined by its opponents. The unity of science is no exception. Consider one recent influential attempt to motivate the appeal that the image of a unified science had for the logical positivists:

> Unity. The very term has always evoked emotions. As a political call to arms, it rouses countries to civil strife, revolution, and international war. The theme of unity is written into the history of the United States, the (former) Soviet Union, and the European nation-states as deeply as any slogan can be. So, one should immediately add, are its antitheses – independence and autonomy. Little surprise, then, if the unification of the sciences, or the autonomy of the sciences, participates in larger cultural debates. In the interwar period, faced with the rise of fascism, the disintegration of the Hapsburg Empire, and the growing tensions between states, a movement grew up under the banner of the Unity of Science. Its roots were various; the motives of its supporters, diverse. But somehow, behind a thinly veiled façade of pure science, there lay a hope and an optimism that the fruits of modernity – the technical wonders of telephone, radio, airplane; the grids of power and railroads; and the scientific spectacles of relativity, quantum physics, and astronomy – could somehow avert calamity. Those who preached the Unity of Science saw inscribed in the new and modern *Lebensform* a rationality that, they hoped, would guard the world against the tide of fanaticism. (Galison 1996: 1)

Although Peter Galison paints a pretty picture of the positivists' motives, it is a *trompe l'oeil*. Notice what he takes to be the *opposite* of unity, namely, independence and autonomy. However, the positivists, being good Kantians (at least in this respect), would have regarded the unity of the sciences as a precondition for science's independence and autonomy. Clearly, Galison, an historian living today, has an understanding of what the previous century was about that differs from the positivists who lived through most of it. In particular, Galison sees the impulse behind nation-building – unificationism's political model – mainly in terms of suppressing local differences to produce a whole that aspires to be greater than the sum of the parts. He supposes that without state intervention, local regions would be normally self-governing. By analogy, then, the question for Galison is whether unification really adds value to sciences that already function well separately.

However, the positivists saw matters in quite the reverse. For them, autonomy emerged from the integration of local regions under a strong national constitution; otherwise, the regions would regress into what Kant first called "heteronomy," whereby they become mere pawns in the hands of larger political and economic forces. Analogously, for the positivists, science must be unified to prevent it from descending into a babble of incommensurable expertises open to the highest (or strongest) bidder. It is no accident that the major modern theorist of international law and architect of Austria's republican constitution, Hans Kelsen, regularly attended the meetings of the Vienna Circle.

Moreover, this difference in the image of the unity of science is not merely of historical interest. A version of the same difference arises in contemporary discussions of science policy (e.g. Gibbons et al. 1994; Nowotny et al. 2000). Here the positivist viewpoint is represented by those who argue that a specific institutional setting – typically the university – is needed to protect the autonomy and independence of science so that it can develop in ways that are both true to its own ends and genuinely serve the public good. This requires, so to speak, a "protected market" in which the sciences are encouraged to interact with each other (through cross-validation checks, interdisciplinary projects, etc.) before facing the larger society. In contrast, Galison's view is represented by those who believe that the sciences are so internally diverse – partly because they already engage society in so many different ways – that any attempt to govern them under a common academic framework is bound to be more coercive than productive. Consequently, the autonomy and independence of the sciences is maintained by allowing them to gravitate to their own "natural markets" or "ecological niches." This latter view, which supports a "disunified" image of the sciences, is often associated with postmodernism, the most influential formulation of which (Lyotard 1983) was expressly written against the viability of the university as a site for knowledge production (Fuller 1999).

2.2. Reducing (away) the philosophical component of reductionism

Contemporary defenses of the disunity of science like Galison's draw their inspiration from the work of the philosopher, Ian Hacking, who came to the disunity thesis as an early Anglophone interpreter of Michel Foucault (e.g. Hacking 1975). From the standpoint of traditional distinction between the "natural" and the "human" sciences, Foucault dealt with practices of institutionalized medicine that straddled the distinction: does "mental illness" refer to a real physiological condition or merely a diagnostic category? The answer turns out to be both, though the condition and category sit together uncomfortably. But Hacking drew a larger lesson. If a discipline's phenomena could be reliably produced without prior commitment to a particular theoretical discourse, the more closely it approximated the practices of the physical sciences (Hacking 1983). On the other hand, the more reliability depended on a prior theoretical commitment, the more the discipline approximated the practices of the human sciences (Hacking 1984).

Hacking (1996) traces modern discussions of the unity of science to the research strategy of *reductionism*. The strategy may be interpreted in two radically different ways, which capture the appeal that it has had for *philosophers* and *scientists*, respectively (in Fuller 2000b: ch. 2, Ernst Mach and Max Planck are contrasted as early 20th-century exemplars of these positions). In terms of Hacking's Foucaultian dualism, philosophers would reduce all sciences to human sciences, whereas scientists would have them reduced to natural sciences. Philosophical interest in reductionism has been spurred by the prospect of a unified language of science, usually the application of logic to elementary observations. This was certainly the program of logical positivism. Here reductionism aimed to level differences in disciplinary discourses that had been misleadingly magnified into ontological ones by experts operating in the public domain. In contrast, scientific interest in reductionism has been less focused on constructing a lingua franca for science than a recipe for arriving at the ultimate constituents of reality. Indeed, the results of this version of reductionism would be expressed in an esoteric language that may even – as in the case of particle physics – refer to objects whose properties defy common sense.

Hacking clearly prefers the scientific to the philosophical attitude to reductionism. Here he claims to follow Leibniz, who identified the quest for the unity of science with fathoming the Divine Plan. Hacking rightly observes that Leibniz was by no means idiosyncratic in his orientation, as the idea of unified science historically descends from a "Book of Nature" presumably written (or dictated) in one language by its author, aka the God of the monotheistic religions. (Not surprisingly, the strongest scientific support today for the version of scientific creationism known as intelligent design theory comes from mathematicians, chemists, engineers, and those biologists who take the code-like character of genes very seriously, who find it easy to imagine God as a big version of themselves (cf. Noble 1997).) At least, this preoccupation with unity cannot be found in areas dominated by the other world religions, including India and China, where many particular scientific and technological advances were made – typically before they appeared in Europe – but without any compulsion to place them under a common theoretical rubric (Fuller 1997: ch. 6).

However, in favoring the scientific over the philosophical attitude to reductionism, Hacking overlooks a salient feature of the Divine Plan that brings the philosophical attitude back into play: if humans have been created to fathom the Divine Plan, then should it not be expressible in a language that *all* humans can understand and apply? Put in sectarian Christian terms, by siding with the scientists over the philosophers on reductionism, Hacking sounds like a Catholic rather than a Protestant. He seems happy to presume that the Book of Nature is written in a language – advanced mathematics and technical jargon – that only scientific experts can understand because they alone have been schooled in the techniques and privy to the sites where the cosmic mysteries are regularly revealed. For the scientific reductionist, so it would seem, the laboratories have become the new seminaries.

These echoes of the religious divisions of the Reformation have shadowed the politics of reductionism since its formal inception in the third quarter of the 19th century. The original reductionists were German scientists who wanted to salvage the unifying spirit of Hegelianism once Hegel's philosophy fell into disrepute for its failure to keep up with French and British developments in the natural sciences. Perhaps unsurprisingly, this first generation of reductionists – including Hermann von Helmholtz, Emil DuBois-Reymond, Rudolf Virchow, and Ernst Haeckel – worked in the life sciences, which even in the mid-19th century were still classed among the "sciences of the spirit," the *Geisteswissenschaften*, and hence well within philosophical jurisdiction (Mendelsohn 1964; Rabinbach 1990: chs 2–5). In today's disciplinary histories, these early reductionists are usually remembered for having introduced specific laboratory or quantitative techniques that helped transform what had been previously "natural history" into a proper experimental science. What is often neglected is the metatheoretical gloss they gave to these techniques. To adopt the techniques of physics was not necessarily to reduce one's science to physics. On the contrary, the early reductionists tended to identify the fundamental ontology of science with "energy," an entity with a dual physical and phenomenological component, rather than "atoms," ultimate units of matter that were theorized as existing below the threshold of human perception.

In Hacking's scheme, Helmholtz, DuBois-Reymond, Virchow, and Haeckel would have to be classed as more philosophical than scientific reductionists. Like the logical positivists a half-century later, they campaigned for an international language of science – and indeed, for science as the basis for an international language of humanity. They regarded the emergence of specialized disciplinary discourses as barriers to the free flow of information and made it their business to promote the public understanding of science. A model for their activities was the *Zollverein*, the customs union that in their lifetime organized the German principalities into a protected market, which in turn removed the economic barriers to Germany's political unification. In this context, reductionism was associated with a certain sense of the "democratization of expertise," namely, the subordination of all scientific authority to the same broad-based standards of evidence and reasoning. Moreover, since the original reductionists themselves came from modest backgrounds and were political liberals (and sometimes even socialists), they also championed entrance examinations in order to open university admissions to a wider range of the population.

Interestingly, among the opponents to their attempt to disseminate the scientific mentality was the physics community itself, which did not appreciate its characteristic methods being applied, and otherwise made available, to all manner of inquiries and inquirers. Indeed, the long-standing hostility of physicists to philosophical reductionism, or its ideological twin "positivism," is one of the most unnoticed features in the modern history of the philosophy of science. Perhaps this is because the anti-positivism of physicists (e.g. Weinberg 1992) has often corresponded to what philosophers call "scientific realism," which is almost identical to Hacking's scientific reductionism.

The renegade physicist Ernst Mach was the key transition figure from the original German reductionists to the logical positivists (Fuller 2000b: ch. 2). He was a contemporary of the former and an inspiration for the latter. Unfortunately, the import of Mach's contribution is often lost because, courtesy of Bertrand Russell, his views were too easily assimilated to those of David Hume and the British empiricist tradition. To be sure, Mach was a kind of empiricist, but he did not share Russell's preoccupation with discovering the "logical atoms" of perception. On the contrary, Mach's own work in experimental psychology studied the construction of meaningful perceptual gestalts from disparate physical stimuli. Mach's focus on the level at which human perception normally occurs was carried over into logical positivism. Even when the positivists spoke of "physicalism," they did *not* mean the reduction of appearances to their physical constituents, such as the motions of atoms. (That way lay Arthur Eddington's notorious paradox of the "two tables" – that is, how the same table can appear solid yet consist mostly of empty space.) Rather, they meant the phenomenology of medium-sized objects, discussion of which could be coordinated through a public language (Creath 1996).

For Rudolf Carnap and Otto Neurath, two positivists who differed on many other points, "physical" was invoked to *oppose* the solipsistic form of empiricism that Russell himself had peddled as "the problem of the external world," which continues to be the central problem of anglophone epistemology. A closer analogue to the positivists' sense of physicalism was the "operationalism" of thermodynamicist Percy Bridgman, who embraced the positivists' unity of science movement when it relocated from Vienna to Harvard during World War II (Holton 1993: ch. 1). Nevertheless, the marginalization of philosophical reductionism from recent history and philosophy of science has been so thorough that it has become difficult even to express the position in an unmisleading fashion. In particular, we have come to associate artificial languages with the displacement of common-sense perception to more esoteric forms of empirical detection, so that the positivist project of a unified language for science appears as more elitist than democratic. Yet, for the positivists, elitism came from supposing that there was one ultimate way of articulating reality that required unconditional deference, be it associated with scientific experts or native speakers of a "natural language."

In other words, the positivists' natural enemy was anyone who believed, following Nietzsche, that "ontology recapitulates philology." This included *both* Martin Heidegger's "jargon of authenticity" and the Oxford school of linguistic analysis that flourished in postwar Britain (Fuller 2002b). The real issue that exercised the logical positivists – say, distinguishing Carnap from Neurath – was whether there had to be any authenticating experiences that corresponded to a statement in the unified language of science or merely a commitment to test the statement in a publicly accessible manner. Gradually, the positivists drifted toward the latter position, which had been always held by a junior and peripheral member of the Vienna Circle, Karl Popper. However, one consequence of this drift was the severance of any connection between the pursuit of science and the adherence to particular beliefs. Science was no longer a vocation but a game. To be sure, this "conventionalism" made science eminently universal, open to anyone – regardless of background beliefs – who is willing to abide by its rules. At the same time, however, it seemed to reduce science to a formal exercise with no psychological purchase on those pursuing it. This led the two great strands of postwar positivism, represented by Carnap and Popper, to move in directions that undermined its original democratic aims: respectively, technical specialization and epistemological skepticism.

2.3. The root image of disunity as intercalation

The long-term effect of the misrecognition of philosophical reductionism has been a certain skew in the narrative told about how the history and philosophy of science got to be as it is today. Galison (1999: 138–44) presents three images that capture the highlights of this narrative, which is allegedly about the search for "foundations" to science. These are depicted in Table 2.2 below. According to the narrative, the positivists originally envisaged foundations as the incontrovertible factual basis on which a succession of theoretical edifices could be built. The antipositivists – represented by the historicist and relativist turn associated with Kuhn (1970) – are then seen as having destabilized this image by stressing that each theoretical edifice (aka paradigm) generates its own unique set of facts. Finally, rather than seeing science as either always stable or always changing, today's historians and philosophers of science plump for an image of science as "intercalated," a metaphor drawn from masonry that is meant to

Table 2.2. Images of the three moments in the modern history of the philosophy of science (after Galison 1999)

Theory$_1$	Theory$_2$	Theory$_3$	Theory$_4$
	Observation		
Time →			

(a) The positivist image of the history of science

Observation$_1$	Observation$_2$	Observation$_3$	Observation$_4$
Theory$_1$	Theory$_2$	Theory$_3$	Theory$_4$
Time →			

(b) The antipositivist image of the history of science

Instrument$_1$		Instrument$_2$		Instrument$_3$	
	Theory$_1$			Theory$_2$	
Experiment$_1$		Experiment$_2$		Experiment$_3$	Experiment$_4$
Time →					

(c) The intercalated image of the history of science

capture the idea that science is not one thing but a set of traditions – of theory, experiment, and instrument – that mutually support each other by undergoing change at different rates. Once this image is granted, then attention turns to the research sites that enable this mutual support through the creation of "trading zones" from which a hybrid language, or "pidgin," emerges.

A sign of the misrecognition conveyed in this account is that Galison's narrative ends just where the positivists would have it begin, since a unified language for science is itself usefully seen as a pidgin developed by inquirers whose various interests intersect in the need to constitute a common reality. Indeed, linguists have identified a process ("creolization") by which pidgins evolve into free-standing languages – such as Latin, Arabic, and Swahili – that eventually supersede the languages used to constitute it (Fuller and Collier 2004: 37–40). And, as I earlier indicated, what ultimately mattered in the positivist quest for a "neutral observation language" was that it could function as a vehicle of communication (of evidence), not representation (of experience). However, whether pidgins evolve into proper languages depends on the removal of barriers to the free intercourse of variously interested scientists. In that case, the incommensurability of paradigms – partly retained by Galison as traditions of scientific practice – is not a mark of cultural identity but a sign of arrested development. My own "social epistemology" starts at this point, namely, that radical conceptual difference is caused by sustained communication breakdown (Fuller 1988).

Moreover, Galison's intercalation model of science recovers an especially pernicious feature of the appeal to traditions in historiography more generally. The clearest precedent in historiographic imagery is the *timeline*, in which a series of geographically distinct cultures are treated in parallel with respect to time. The first notable appearance of the timeline was in the second edition of *Encyclopaedia Britannica* (1780), in a newly expanded article on "History" by the Scottish Enlightenment philosopher, Adam Ferguson, an early purveyor of the idea that societies are self-organizing systems. Ferguson's timeline was organized orthogonally to Galison's. Thus, the multiple layers of traditions appeared as a series of adjoining columns, and time was marked by a horizontal, rather than a vertical, line that cut across all the traditions. (Interestingly, Ferguson's timeline began at 4004 BC, the origin of Creation implied by the ages of the biblical patriarchs.)

To his credit, Ferguson recognized the implications of the image, namely, that "revolutions" (his term) are caused by the interaction of peoples with inherently different characters. Ferguson did not mean to cast doubts on this interaction, but simply to explain its tendency to be catalytic. Here he was informed by a viewpoint that by the next generation would be called "Lamarckian," namely the communicability of acquired traits to the next generation, which "evolves" by building on their collective intelligence. This patterned sequence of organisms constitutes that bio-social duplex known as "culture" or "race." Writing a century before the rise of modern, Mendelian genetics, Ferguson was more inclined to stress the emergent features of this duplex than their composition from traits that could be combined with others, in the right environment, to form different cultures or races. In today's shorthand, we would say that Ferguson "essentialized" the divisions presented by the history of humanity. So too, though with less explosive political consequences, Galison essentializes the different traditions of science. Thus, whereas Ferguson neglected the role of diffusion and migration, Galison occludes the overlap in the people who have contributed to the various intercalated traditions.

3. Unity and Disunity as Expressions of Constructivism and Realism

Because disunificationists define the terms for discussing the unity of science, the issue has lost much of its centrality in contemporary philosophy of science. In sections 3.1 and 3.2, I relate the dormant unity–disunity distinction to the much more active debate between *constructivism* and *realism*. Generally speaking, the unificationsts of most concern here (i.e. the positivists, or philosophical reductionists) introduce a distinctly normative perspective on knowledge that is designed to counteract knowledge's natural tendencies toward dispersion, primarily by recognizing the constructed character of the distinction between evaluating and applying knowledge claims. In that respect, the unity of science is clearly a "project" that requires a deliberate strategy, which is sketched in section 4. Before that, I examine the constructed character of the natural-normative and the evaluation-application distinctions.

3.1. The natural and the normative: aligned or opposed?

Recalling section 2.2, Hacking's contrast between scientific and philosophical forms of reductionism is a version of a distinction that contemporary philosophers of science officially find more significant: *realism vs constructivism*. Generally speaking, scientific reductionism is a form of realism, and philosophical reductionism a form of constructivism. For our purposes, it is important to note that this distinction is orthogonal to the difference between unity and disunity. Thus, it would be a mistake to assume that, say, most unificationists have been realists or vice versa. Of course, some unificationists have been realists (e.g. Wilson 1998, which is discussed in 2.1 above). But insofar as the unity of science has been something that philosophers have proposed as a program for organizing the sciences, the assumption has been that, left to their own devices, the sciences tend toward dispersion and fragmentation, which is to say, disunity (Dupré 1993). In other words, unless explicit efforts are taken to realign the sciences, the force of some "external reality" on them is not sufficient to ensure their ultimate unification. The logical positivists are thus probably best seen, on balance, as *constructivist unificationists* who were keen on bringing order to natural disorder. In contrast, much of today's antipositivist sentiment is motivated by a *realist disunificationism*, which owes equally to Kuhn's (1970) paradigm-based relativism and Lyotard's (1983) narrative-based postmodernism: it would permit, so to speak, the thousand flowers that have always already been blooming.

When considering the relationship between the unity–disunity and the realism–constructivism distinctions, a key background distinction is between *the natural* and *the normative*, specifically, whether these two categories are supposed to be aligned or opposed. The ideal case of alignment is where one claims to infer what is normative from what is natural (or "what ought to be" from "what is"). This view is common to a wide range of views including both attempts to derive a "revealed history of humanity" and an "evolutionary ethics." The ideal case of opposition is where one defines the normative as explicit resistance to the natural. Once again, a diverse range of views would fall under this rubric, ranging from existentialist assertions of human freedom to legal sanctions designed to alter citizens' default behaviors. In the two contexts, "natural" has somewhat different connotations. "The natural" that is aligned to the normative usually implies a foundation in some underlying reality, whereas "the natural" that is opposed by the normative is a statistical regularity that is itself an incomplete or distorted expression of reality's potential.

Thus, realist disunificationists take the natural and the normative to be aligned and constructivist unificationists take them to be opposed. This difference is most clearly reflected in attitudes toward the work of philosophy. Realist disunificationists see no role for philosophy as an independent discipline lording over the sciences. It is here that the later Wittgenstein's calls for philosophy to "leave the world alone" make the most sense. In contrast, constructivist unificationists see a large role for philosophy in manufacturing metalanguages and other cross-disciplinary bridges.

3.2. Evaluation and application: clear or blurred?

However, matters turn out to be more complicated because realism and constructivism as general epistemologies differ substantially in their views about the hardness of the line that separates what may be alternatively called "subject vs object," "signifier vs signified," or "mind vs world." The realist draws a sharp, and the constructivist a blurred, line between the two sides of each of these binaries. In practice this means that compared with constructivists, realists have a much clearer sense of the situations in which science is and is not being done. Thus, "realist disunificationism" presupposes, as we saw in Galison's political image, that the individual sciences are already self-governing entities, each assigned to its domain of reality. In that sense, "realists" do not normally resort to "common sense" for their sense of reality; rather, they are much closer in spirit to Platonism, Cartesianism, and other philosophies that presuppose a stratified view of reality.

To be sure, realists vary over the best way to characterize and access the "surface" and "depth" of reality – and some recent disunifica-tionists like Hacking envisage a more horizontally arranged ontologi-cal space that presupposes a metaphorical distinction between "core" and "periphery." Nevertheless, the "deep" or "core" end of reality – or "fundamental ontology" – is invariably a restricted realm that requires special methods and training. From the realist point of view, the constructivist appears to confuse surface and depth, appearance and reality, belief and knowledge, verification and truth, and so on. This is because constructivists hold that what counts as the "surface" or "depth" of reality is itself a construction of the knowledge system, not a representation of something that exists outside the system. In that sense, the constructivist collapses the distinction between the philosophical disciplines of epistemology and ontology, which the realist holds sacrosanct.

For their part, "constructivists" are often mischaracterized as rela-tivists and even nihilists who believe that each culture (or person) is entitled to their own truth (Fuller 2006a: ch. 2). In this misreading, the verb "to construct" means "to make up out of nothing," as if con-structivists were arguing for a view of reality that has no basis outside itself. A more correct interpretation of "to construct" is "to make up out of something," a phrase that brings out better reality's processual character as the past provides the raw material out of which the future is made. Consequently, constructivists blur philosophical distinctions that realists want sharpened. In particular, realists draw a strong dis-tinction between the contexts of *evaluating* and *applying* knowledge claims, whereas constructivists treat the two contexts as the same, often under the rubric of *knowledge production*.

The basic idea here is that realists hold that knowledge claims first need to pass a quality control check in a cloistered expert setting – which may be a laboratory, a circle of peers, or even one's own secure mental space – before those claims are unleashed on the world. Only in the expert setting can one identify the salient variables, underlying causes, intuited essences, and so on. The real is thereby made appar-ent. However, if the world ends up worse as a result of applying knowl-edge of this reality, then that will be blamed on the low intellects or corrupt morals of the appliers. In contrast, constructivists do not sharply distinguish between how professional scientists and the rest of humanity access reality. Such public spaces as the agora, the battle-field, or the hospital are equally good as sites to manifest knowledge claims as more cloistered settings (Nowotny et al. 2000). Thus, evaluation and application collapse into each other, and one is never really sure (nor perhaps should one be) whether a given outcome reflects human will or objective reality. "Knowledge" here turns out to be an irreducible mixture of the two.

In terms of social epistemology, realism may function ideologically to absolve scientists from responsibility for the consequences of actions taken on the basis of their authority but without their involvement. The realist holds that the responsibility of scientists ends with the consequences logically entailed by their theories and empirically predicted in their laboratories. However, to avoid the moral quandary in this way is to conflate the *intended* and the *anticipated* consequences of one's actions. For example, Einstein and Bohr did not intend for their physics to result in the atomic bomb. But at some point in the history of atomic physics they could have anticipated that the theory would be put to that use: maybe not in 1910, but certainly by 1945, when the US dropped the atomic bomb on Japan. Once some disturbing precedents are set, there is little excuse to allow those precedents to become normal practice.

Unfortunately, the history is not so philosophically straightforward, since Einstein actually urged Franklin Roosevelt to start the US atomic bomb project. Of course, Einstein and other atomic physicists eventually took responsibility by campaigning against nuclear weapons, once they started to proliferate. Those who saw this activity as part of their responsibility *as scientists* would be classed as constructivists. In contrast, realists tend to make the idea of "unintended consequences" do too much work, as if we could only learn from our mistakes in our cloistered settings but never in the world outside the cloister. Accordingly, from a constructivist standpoint, scientists are often irresponsible because they pretend to know much *less* than they do – or could know, if they devoted as much effort to understanding the potential applications of their research as to how their peers will evaluate it. However, realism may discourage precisely this moral impulse because of its strong distinction between the contexts of evaluating and applying knowledge claims.

One of the most historically sustained examples of the realist–constructivist sensibility is the difference between biomedical researchers and practicing physicians. To be sure, this difference is obscured by the emotively charged question: would you allow a realist or a constructivist treat you for cancer? Yet a realist engaged in biomedical research would probably resist making *any* substantial interventions until the underlying causes are established beyond a reasonable doubt. Indeed, the rhetoric surrounding the people who doubt the connection between cigarettes and lung cancer, HIV and AIDS, and other such attempts to link cause and effect, is usually very "high realist." In complete accuracy, if not always sincerity, these people warn that we should not mistake correlation for causation. Thus, an excuse is provided for protracting basic research programs and perhaps even multiplying the avenues of research pursued, without doing anything for the people currently suffering from lung cancer or AIDS.

In contrast, practicing physicians who advocate a substantial invasion of people's bodies do not usually think like a pure realist. They are more inclined to blur the contexts of testing and applying knowledge claims. If the patient lives after, say, having undergone chemotherapy, then chemotherapy is "constructed" as the decisive cause; if the patient dies, then the cancer is constructed as the cause that resisted the treatment. However, the outcome cannot be predicted with certainty because there is no necessary causal grounding for it. It is by going through the treatment that the knowledge claim in question ("chemotherapy cures cancer") is simultaneously tested and applied. In this sense, physicians are practicing constructivists, which is why so much of what they do is tied up with legal and moral issues.

In the history of the West, realism has been a revanchist move that recurs periodically whenever constructivist-inspired activities have had palpably bad effects on society, and society has responded by threatening to curtail free inquiry altogether. In psychoanalytic terms, realism is the "defense mechanism" of the "once burnt, twice shy" inquirer. Put somewhat less charitably, realists try to have their cake and eat it by taking responsibility for the good consequences of their research but refusing responsibility for the bad ones. From that standpoint, constructivists can at least claim the virtue of symmetry, a willingness to take responsibility for *everything* they do.

Thus, an encapsulated history of Western knowledge production would start by noting that Plato's founding of the Academy in the outskirts of Athens was an explicit attempt to shelter inquiry from the politically destabilizing effects of the public exercise of reason by the Sophists, which had culminated in the Peloponnesian Wars. This resulted in the creation of an environment fit for contemplating the ultimate forms of reality. Two thousand years later, the Thirty Years War in Europe over alternative interpretations of the Bible resulted in such seminal scientific institutions as the Royal Society of London and L'Académie des Sciences, which were chartered as ideologically neutral zones for the pursuit of knowledge. Another three centuries later, the various idealistic and positivistic schemes for insulating the development of knowledge from other societal developments – Hegel's and Comte's most notably – were once again attempts to contain and channel "Enlightenment" impulses which had originally ended in the bloodshed of the French Revolution. This pattern continued in the 20th century, with logical positivism and Kuhn's paradigm-based theory of scientific change featuring as increasingly sophisticated attempts to protect science from external interference, especially given, on the one hand, the Weimar backlash against science following Germany's defeat in World War I and, on the other, increasing public concern in the US about the application of scientific research in the wake of Hiroshima (cf. Sassower 1997).

4. Conclusion: Towards Recovering the Unity of Science

We live in the time when the allegedly discredited project of unifying the sciences is being given a "decent burial" by those who regard it as a misguided precursor of what they are now themselves doing. That many of today's disunifcationist apologists for logical positivism – Nancy Cartwright, Peter Galison, Michael Friedman – are distinguished historians and philosophers of *physics*, like the positivists themselves, is telling, since physics has been clearly the discipline that has both pushed the unificationist agenda hardest during the Cold War and suffered the greatest loss in prestige and funding in its aftermath. The logician Quine liked to speak of a "principle of charity" that underwrites our understanding of other times and places, but the charity extended to the positivists by today's disunifiers is a backhanded compliment that deserves its own name: the *principle of magnanimity*, whereby our ability to excuse those who differ from us is taken as an indirect measure of our own success. Thus, the logical positivists are nowadays praised, not for their unificationist ambitions but for their honest recognition of the plurality of scientific disciplines that continually confounded those ambitions. Their honesty matters because "we" are now disunificationists (Cat, Cartwright, and Chang 1996). While such an interpretive stance would have provided cold comfort to the positivists, at least it has led to a renewed interest in fathoming their original motives. However, lost in this particular transaction between past and present is any future hope for unifying the sciences.

Nevertheless, as this chapter has endeavored to show, unification-ism has not traditionally had to rely on the institutional vicissitudes of physics. The *university* provided the rhetorical space for unifying knowledge on a regular basis, as the demands of a common curricu-lum compelled the practitioners of different disciplines to integrate their disparate inquiries together. These demands simultaneously enabled academics to escape their narrow guild interests and prepare the groundwork for a public that would be receptive to academic work, namely, the future leaders that make up the student body. However, this strategy is now being eroded by the piecemeal and vocational character of academic instruction, a result of the increas-ing use of credentials as a principle of social stratification. Meanwhile, the current generation of academics will spend at least some time as teachers and researchers on fixed-term contracts, and some will spend their entire careers that way. This provides little incentive for younger academics to invest in the institutional future of the uni-versity, and indeed relatively few of them are inclined or encouraged to participate in peer review processes, publish substantial synthetic works, or occupy academic leadership positions – activities that cannot be justified mainly in terms of short-term career advantage.

We have seen how contemporary historians and philosophers mis-recognize the motives and metaphors behind unificationism. A final striking case of this misrecognition is worth noting because of its recent revival of biological accounts of the human condition, which we shall repeatedly encounter in these pages. The image of the Swiss Army Knife originally introduced by Rudolf Carnap as a model of the human mind (Margalit 1986; Fuller 1993: 142) was intended to motivate the positivist search for a unified inductive logic that would codify the principles of empirical reasoning. However, inductive infer-ence is inherently fallible because it attempts to anticipate the future based on data from the past, which are themselves often imperfect, due to a variety of constraints on our mental and physical resources. Nevertheless, we do rather well overall with this faulty logic, much like a Swiss Army Knife whose blades are suboptimal for each of its tasks but good enough for all of them.

In recent years, this image has been turned on its head by evolu-tionary psychologists, who may be understood as interested in reading the disunificationist image of science into the structure of the human brain as an adaptive feature of natural selection. Accordingly, now the brain is supposed to be like a Swiss Army Knife because it consists of separate domain-specific modules (e.g. a part or process of the brain devoted to perceiving certain shapes), which are like the blades that do each of its jobs well with little more in common than physical co-location in the same piece of metal (Cosmides and Tooby 1994: 60). The question then becomes how the brain's modularity could be a product of evolution rather than design (as in the case of the Swiss Army Knife). I would speculate that these two radically different interpretations of the same image are made possible because the positivists considered the Swiss Army Knife as a functioning tool, whereas the evolutionary psychologists regard it an idealized object.

So, how might we reverse the ideological dominance that the disunificationists exert over the interpretation of the unity of science project? I propose a two-pronged strategy.

First, I would draw attention to the historical resonance of the word "encyclopedia" that figured so prominently in logical positivist projects. While today's disunificationists have been keen to stress the positivist interest in building cross-disciplinary bridges through systematic cross-referencing, they have downplayed the key Enlightenment feature of the encyclopedia that made it so appropriate for the positivists' unifying impulses. Specifically, the cross-referenced articles in Diderot's original *L'Encyclopédie* often contradicted each other, forcing readers – typically non-specialists in any of the referenced fields – to resolve the different authorities for themselves. Indeed, it was just this open display of expert disagreement that had made *L'Encyclopédie* such a controversial commercial venture: suddenly, received wisdom had been turned into fodder for public debate and perhaps even an opportunity for a higher-order dialectical synthesis that would supplant the old expert categories. (This was, of course, the attraction of the encyclopedia format to the German idealists, most notably Hegel.) Similarly, the encyclopedism of the logical positivists was not meant to foster the temporary cross-disciplinary collaborations favored by today's disunificationists, but rather an ongoing collective effort to transcend the barriers that result whenever knowledge is concentrated as a source of power.

This brings me to my second ideological proposal for reviving unity's fortunes. In section 1.1, I mentioned that a popular inductivist approach to unifying the sciences has been to imagine separate disciplines as tributaries flowing into one river that then gives them a common focus and direction. The positivists used this image not only for the unification of the sciences but for the unification of individual discoveries under one science. Here the distinction between contexts of discovery and justification was introduced (Fuller 2000b: ch. 1). However, this use of the fluvial image – a river and its tributaries – makes the unification of science appear to be an attempt to homogenize differences and consolidate power. As we saw in the quote from Galison in section 2.1, this in turn makes it easy rhetorically to endorse the disunity of the sciences as natural. However, a river may also have a delta that forms as its waters empty into a larger body of water. The image of the delta, with its focus on the distribution of the river's waters to sustain a variety of shifting and intersecting life activities, is the image of unification worth promoting, and one more consonant with the original encyclopedic ambitions of the logical positivists, who promoted unity as such a vivid image for science. Some implications of that image for the issues raised here are provided in Table 2.3.

Table 2.3. Alternative images of the evolution of science

Heyday of the image	19th–20th century	20th–21st century?
Metaphor guiding the distinction	Tributaries flowing into a major river	A major river flowing into a delta
Prima facie status of discovery	Disadvantage (because of unexpected origins)	Advantage (because of expected origins)
Ultimate role of justification	Concentrate knowledge through logical assimilation	Distribute knowledge through local accommodation
Background assumption	Discoveries challenge the dominant paradigm unless they are assimilated to it	Discoveries reinforce the dominant paradigm unless they are accommodated to local contexts
Point of the distinction	Turn knowledge into power (Magnify cumulative advantage)	Divest knowledge of power (Diminish cumulative advantage)
Nature of reduction language	Esoteric	Exoteric
Textual image of nature	Single-authored book	Multi-authored encyclopedia

Part II

The Democratization Problem

3

Contrasting Visions

It is one thing to demarcate science from non-science philosophically. It is quite another to govern how what lies inside the demarcated boundary relates to what lies outside it. The former problem is formal, the latter substantive. In this chapter, we shall consider two rather different approaches to this latter problem, both of which travel under the banner of "democratizing science." The first is that of Bruno Latour, the leading figure in STS today, whose main democratizing move is to level differences in the kinds of entities that are eligible for representation in what he calls the "parliament of things" (Latour 1993). The second half is my own effort, which is informed more explicitly by the history of civic republican democracy, which has specialized in designing institutions of conflict management. (Those interested in witnessing a direct encounter between our two positions should consult the transcript of our 2002 Hong Kong debate: Barron 2003.) However, because Latour takes an especially metaphysical approach to politics, we begin with a survey of what I call "the two grand metaphysical strategies" in the history of Western thought on which he draws, which in turn ties in with some of the metascientific issues concerning demarcation raised in Part I.

1. The Two Grand Metaphysical Strategies: Instantiation and Emergence

The earliest precedents for discussions of knowledge integration can be found in ancient metaphysics. A metaphysics is a scheme for organizing reality that is constructed by relating two dimensions: an apparent (folk) dimension that distinguishes particular "individuals" from the "properties" they share with other individuals and a real (scientific) dimension that distinguishes "abstract" terms from the "concrete" objects in which they can be realized. The fundamental question for any metaphysical scheme is how to map the first dimension onto the second: which apparent objects are abstract and which are concrete? Except in extreme forms of Platonism, abstract objects typically cannot exist unless materially realized in some concrete objects, which, in turn, place constraints on the ways in which the abstract objects can appear. By themselves, abstract objects are indeterminate. And so, while abstract objects enjoy the metaphysical virtue of primitiveness, they pay the price of never existing in a pure form but only as part of a mixture with other such primitives.

There are two general strategies for relating individuals and properties to abstract and concrete objects: a *standard* and a *non-standard* strategy. The standard strategy focuses on *instantiation*, the non-standard on *emergence*. The instantiationist takes properties to be abstract and individuals to be concrete, whereas the emergentist treats individuals as abstract and properties as concrete. Table 3.1 provides a guide to what follows.

The pre-Socratic Greek philosophers set the terms of the most fundamental debate, though their explicit concern was not with the integration of knowledge but the constitution of reality. By the Christian era, the theological difference made by the distinction was clear: the instantiationist God creates according to a preordained plan, whereas from the emergentist God comes half-formed creatures endowed with

Table 3.1. Précis of the two grand metaphysical strategies

Grand strategy	Standard	Non-standard
Metaphysical move	Reducing the many to the one (instantiation)	Realizing the part in the whole (emergence)
Image of reality	Reality is planned	Reality emerges
Activity focus	Intellectual consumption	Intellectual production
Value of science	Science saves labor (epistemologically, by explaining the most by the least)	Science adds value (ontologically, by bringing new entities into existence)
Nature of definition	Functional equivalence (paradigmatic)	Contextual delimitation (syntagmatic)
Model of generalization	Universalization	Globalization

the freedom to complete themselves. In the one case, humans have a fixed essence; in the other, their essence is defined precisely by its unfixed character. Someone who believes that rather different things – humans, animals, computers – can possess a "mind" in the same sense of the term by sharing certain formal properties is probably an instantiationist; whereas someone who defines mentality in terms of the presence of certain physical conditions – say, a threshold of neural complexity – is probably an emergentist. This contrast can be found even in modern evolutionary biology. Notwithstanding his avowed atheism, Richard Dawkins's (1976) notorious "gene's eye-view of the world" is instantiationist because he locates the motor of biological evolution in a gene-based drive to self-reproduction, whereas the emergentist would argue that higher-level interactions (say, between the individual organisms carrying genes) do more to determine the overall direction taken by evolution (Bechtel 1986).

Characteristic of instantiationists in the history of metaphysics is that individuals are said to be composed of a combination of underlying properties. Thus, medieval scholastics spoke of resolving a particular into its component universals, or 20th-century analytic philosophers declared that a proper name is coextensive with, or its meaning is exhausted by, a "definite description," that is, a finite conjunction of predicates. Given the abstract nature of these properties, the individual is often portrayed as the spatio-temporal point at which the properties converge. Without spatio-temporal moorings, abstract objects would be "indeterminate" in the sense of *unbounded*. In contrast, the emergence strategy regards each property as what Hegelians call a "concrete universal" that consists of individuals organized in a distinct way. For example, political theorists have periodically spoken of each person as literally part of a "body politic" or a "social organism." Accordingly, to be a "human" is *not* to possess a property that every other human being has individually; rather, it is to possess a property that is jointly realized with other individuals interacting in an appropriate fashion. Without this interaction, the identity of each individual human would be "indeterminate," in the sense of *incomplete*.

The classic metaphysical conundrum for the instantiationist is *the one and the many*: how can the same property, say, "humanity," belong to an indefinite number of individuals? Why does a property not diminish as it participates in the definition of each new individual, rather than enhance its ontological status, which is to say, increase its claim to reality? These instantiationist questions are implicitly answered in legal systems that accord individuals just enough freedom to enable everyone to enjoy the same amount of freedom. In contrast, the emergentist regards "being" as completely vacuous, since the term fails to distinguish among entities. Informing this judgment is that the truly deep metaphysical problem is *the part and the whole*: how can the activities of spatio-temporally separate individuals be arranged so as to enable the emergence of some higher order unity? Given the inherently partial nature of individuals, why do they not constantly interfere with or simply ignore each other, instead of interacting in a fashion not only mutually beneficial but also "good" in a sense that transcends their aggregated interests?

In the history of legal thought, the difference between the instantiationist and emergentist is best seen in contrasting English and German attitudes towards "freedom": the former is conceptualized as unrestricted action to which all citizens are equally entitled (on the model of the Roman *libertas*), whereas, in the latter case, freedom emerges through the enactment of a system of rights and duties that circumscribe the sphere of action for different citizen classes (as literally suggested by *Freiheit*, "freedom" as the domain within which one is free). To the English lawyer, German "freedom" looks like a rationalized version of feudalism; to the German lawyer, English "freedom" looks like what Emile Durkheim (1952) called "anomie," lawlessness. Isaiah Berlin (1969) canonized the instantiationist and emergentist conceptions of freedom as, respectively, the "negative" and "positive" concepts of liberty.

The sets of questions suggested by these two metaphysical strategies – *the one vs the many* and *the part vs the whole* – imply two radically different conceptions of inquiry. Put in terms an economist could appreciate, the instantiation strategy defends itself as a more efficient way of carrying out the emergence strategy, whereas the emergence strategy responds by revealing the hidden labor costs of the instantiation strategy. According to the instantiation strategy, inquiry is an intensive, perhaps even microscopic, search for the essential properties into which an individual can be analyzed without remainder and which together can be used to synthesize the individual without cost. By contrast, the emergence strategy works by differentiating a whole into its proper parts – but at a cost, since it is not clear that the process can be reversed, so as to allow the parts to be reintegrated into the original whole.

Finally, in terms of structural linguistics, instantiationism and emergentism are manifested in the distinction between the *paradigmatic* and the *syntagmatic* dimension of definition. In both pairs, the former term defines the meaning of a word or phrase by substituting synonymous expressions (e.g. a definite description), whereas the latter term defines by functional differentiation in a system, as in the context of a narrative (Culler 1975). A general theory of semantics relates to language paradigmatically, while the hermeneutics of a specific text relates to language syntagmatically. Similarly, the "old" sociology of science represented by, say, Robert Merton and Joseph Ben-David defines the scientist in instantiationist terms as someone who satisfies a complex social role, while the "new" sociology of science represented by, say, Bruno Latour and Steve Woolgar regard the scientist as a specific agent who emerges from within an actor-network (Latour 1992; Fuller 1996b, 2000b: 365–78).

1.1. These strategies applied to knowledge integration: universalism vs globalism

The classical philosophical debate over the constitution of reality mirrors today's concerns about the relationship between the individual knower and the collective body of knowledge. Is it one–many or part–whole? In other words, should the individual be seen as one of many similarly equipped knowers whose similarity marks them as members of one community, or as the possessor of a unique knowledge base that complements the unique knowledge possessed by the community's other members? This question recalls Emile Durkheim's (1997) original sociological formulation of the distinction between instantiationist and emergentist metaphysical strategies in terms of "mechanical" and "organic" solidarity as successive stages in the evolution of the division of labor. Indeed, one explanation for the rise of modern science in an otherwise economically backward and politically disorganized Western Europe can be cast in these terms. Unlike the great Eastern empires, and even ancient Greece and Rome, class distinctions could not be rigidly enforced, which allowed a sense of organic solidarity to emerge from the free exchange of academics and craftsmen, as institutionalized in the experimental testing of theoretical hypotheses on specially designed machinery (Fuller 1997: ch. 5).

In contemporary discussions of the organization of knowledge, the instantiation-emergence distinction appears most clearly in, respectively, *universalist* vs *globalist* knowledge policy strategies. The former aims for law-like regularities that apply in all societies, whereas the latter aims for a unique narrative that accounts for the one world-system in terms of relationships among its constituent social formations. Before considering the implications of this contrast for knowledge policy as such, it is worth examining the vivid versions of universalism and globalism that may be found in both Marxist and capitalist accounts of economic history.

In terms of Marxism, Lenin followed Marx's own practice of treating his theory as a transnationally repeatable blueprint for economic change, whereby leaders like Russia would show the rest of the world the way to the proletarian revolution. In contrast, Trotsky drew on Marx's Hegelian roots to hold that there is no such blueprint, only a gradually emergent global process, which therefore renders nonsensical the idea of socialism in one country. An updated version of such globalist Marxism is Manuel Castells's (1996–8) recent characterization of the contemporary world as a "network society." By this expression, he does not mean the ascendancy of information technology as a mode of production that recurs across many nations. That would be congruent with a universalist perspective. Rather, Castells means the transnationally variable ways in which information technology has reconned the entire world-system. This includes deliberate backlashes against computer networks, the use of computers in unintended and perverse ways, as well as the unwanted disparities in wealth that the networks have produced at a global level.

In terms of capitalism, consider, on the one hand, Walt Rostow's (1960) stage-based model of economic growth; on the other, Alexander Gerschenkron's (1962) thesis on the relative advantage of backwardness. The former is universalist and the latter globalist. Rostow's "non-communist manifesto" followed Marx in believing that the path to economic progress is indefinitely reproducible, whereas Gerschenkron grounded his own anti-Marxist stance on the fact that the economic success of one nation may serve to prevent other nations subsequently succeeding by the same means. Thus, whereas Rostow saw latecomers to capitalism as more efficiently repeating the same stages as their predecessors, Gerschenkron cast the latecomers as innovators forced to overcome the phenomenon of "path-dependent" development. The growth of capitalism in Japan reveals interesting differences in emphasis between the two approaches. A universalist would stress how the Japanese overcame cultural differences with the West to embark on accelerated capital accumulation, while the globalist would focus on the ways they capitalized on those very differences.

The contrasting accounts offered by universalist and globalist economic history suggest that these two perspectives can be distinguished by the effect of scale and scope on social relations. According to the universalist conception, societies can expand indefinitely without changing their fundamental nature, and many societies can share the same fundamental nature, often by one imitating another. Constraints, such as there are, come from the outside, most crudely, as a selection environment that curtails population growth. When the mass exportation of free markets, a technological innovation, or a scientific paradigm is treated unproblematically, the universalist mentality is usually at work. Unsurprisingly, the failure of these foreign exports to be integrated in native environments is usually described in negative terms, such as ideologically inspired "local resistance" to something that *ought* to be universally available. In contrast, according to the globalist conception, an expansion or contraction of the parts necessarily alters their interrelation, which in turn changes the nature of the whole that the parts constitute. This throws into doubt the idea that either individuals or societies can ever simply "imitate" each other: something is both lost and gained in the process. Predecessors either crowd out successors or unwittingly instruct them on how to improve on their achievement. As economists might put it, the universalist conception "exogenizes" changes in scale and scope, whereas the globalist conception "endogenizes" it.

In terms of knowledge policy, universalism and globalism are expressed, respectively, as *interdisciplinarity* and *transdisciplinarity* (Klein 1990; Gibbons et al. 1994). Interdisciplinarity presupposes the existence of disciplines that between them carve up reality into distinct domains of inquiry, each governed by laws, which in some combination can be used to provide richer understandings of a particular phenomenon that are taken to be an instantiation of those laws.

In contrast, transdisciplinarity presupposes that reality escapes any combination of disciplinary perspectives, which are themselves treated as little more than an artifact of the last 150 years of the history of the Euro-American university system. Thus, in the case of tropical disease, one may adopt either an interdisciplinary approach that brings together specialists from biomedical and environmental science and public health policy, or a transdisciplinary approach that treats tropical disease as a domain of scientific inquiry in its own right, requiring expertise that is not reducible to a combination of existing disciplinary practices.

This example epitomizes the problems facing academic administrators and research managers in the "periphery" of the world's knowledge-system: is it best to attempt to reproduce "core" Western research institutions in the periphery, or to develop alternative and perhaps complementary institutions that succeed on their own terms? The dilemma is acute because the world's knowledge-system now seems to be constituted so as to make it marginally more advantageous for peripheral knowledge producers to imitate, however unsuccessfully, core research trajectories than to innovate native ones (Shahidullah 1991). A key indicator here is the *Science Citation Index*, which is more likely to include peripheral knowledge producers who publish in core journals than in peripheral ones.

1.2. The postmodern condition as a challenge to these strategies

In the 19th century, instantiationism and emergentism differed along the inanimate–animate divide. At that time, instantiationism was associated with Newtonian mechanics and Platonic metaphysics. Emergentism corresponded to vitalist biology and Aristotelian metaphysics. However, since the 20th century, this divide has come to be blurred, as theories on the instantiationist side of the divide have mutated and migrated to the emergentist side, producing the *postmodern condition* diagnosed by, among others, Jean-François Lyotard (1983). This development captures most of the dominant movements in 20th-century science, all of which have emphasized the "irreversibility" of temporal change but have stopped short of conferring purposefulness on the emergent direction of change. Examples include the neo-Darwinian theory of evolution, dissipative structures in thermodynamics, indeterminist interpretations of quantum mechanics, catastrophist mathematics, and chaos and complexity theory.

The general explanatory framework common to these theories emboldened the Gulbenkian Commission convened by Immanuel

Wallerstein (1996) to call for a radical transformation of the social sciences. The commission observed that these theories explain irreversible change, roughly, in terms of the effects of a local disturbance reverberating throughout an entire system. At the very least, phenomena conforming to this pattern challenge a methodological dictum common to Aristotle and Newton, namely, *the proportionality of cause and effect*. Put another way, the postmodern condition in science highlights the distinction between *propagation* and *repetition*: specifically, the former need not imply the latter. This is not because the effect might represent a corrupt version of the cause (as the instantiationist might think). Rather, it is because the medium by which a cause produces its effect might be itself causally relevant. A case in point is the active role played by sexual reproduction in inheritance. Gilles Deleuze (1984) has been the philosopher most sensitive to this point.

In the case of sexually reproducing organisms, a parent passes its genetic material to an offspring without guaranteeing that the offspring will be identical to itself. This is due not only to the requirement of two parents, but the statistical nature of the laws of heredity. Thus, any given parent reproduces without strictly repeating. Therefore, it would be misleading to say that two generations of organism are simply "instantiations" of the same species. Yet it equally does not follow that the new organism constitutes an improvement over its parents, as modern versions of the emergentist strategy have often supposed (for example, Lamarckianism and Hegelianism). In evolutionary biology, this "non-emergent non-instantiation" is explained by distinguishing between two statistically interrelated levels: *how* genes are selected (namely, an organism's manifest capacities and the conditions under which they are exercised) and *what* is selected (namely, the range of capacities expressible by the organism's offspring). The relevant jargon is "phenotype" (or, in Richard Dawkins's (1982) sense, "extended phenotype") vs "genotype."

To be sure, there is precedent in the history of the human sciences for the disproportionality of cause and effect: so-called "invisible hand" accounts of the emergence of a stable social order as the unintended consequence of aggregated self-interested actions. However, the Gulbenkian Commission invoked a more negative, Marx-inspired interpretation of the invisible hand metaphor. Whereas the Scottish Enlightenment originators of the metaphor – such as Adam Smith and Adam Ferguson – tended to envisage a country *benefiting* from the invisible hand at work, the 20th-century commission treats the entire globe as a system that, on the whole, *suffers* from what are essentially accidents of history coming to be treated as laws of nature, simply on the basis of their persistence. Wallerstein's own world-system theory most explicitly develops this point, as suboptimal local patterns of production and trade are said to have forced medieval Europeans to embark on an expansionist campaign that eventuated in the modes of world domination characteristic of the modern era.

1.3. From instantiation and emergence to closed and open sciences

Although the postmodern condition has blurred the instantiation–emergence distinction, the Gulbenkian Commission has resurrected the distinction in terms of, respectively, a *closed* from an *open* conception of science. Specifically, an instantiationist metaphysics lends itself to a closed social science, in the sense that the social world – much like the physical world in Newtonian mechanics – is portrayed as "closed" under a set of laws that apply for all space and time. In contrast, an emergentist metaphysics lends itself to an open social science, because the social world is portrayed as continually generating novel consequences from the interaction of known tendencies, including those associated with the inquirer's frame of reference.

As was observed in section 1, even a system of laws that seems to contain a strong developmental component can be said to be "closed" in the relevant sense. Take the stages in Marx's dialectical materialist account of history, Kuhn's (1970) stages of scientific change through normal and revolutionary phases, or Jean Piaget's (1976) stages of cognitive development in the child. All of these accounts are universal in scope, and hence apply in case after case. In particular, the laws described in these accounts remain unchanged by the number or kind of cases to which they are applied. There is no feedback from the applications to the laws. Thus, the transition from capitalism to socialism is supposed to be the same regardless of the country's specific history; the phases in the growth of scientific knowledge is the same regardless of discipline; the transition from concrete to abstract operations the same regardless of the child's gender or birth order.

In contrast, in an open conception of social science, there are several respects in which the inquirer participates in constituting the objects of inquiry. The most obvious ones concern the inquirer's background value commitments, but no less important are more objective features of the inquirer's location in space and time. From the standpoint of an open social science, the status of Marx's laws of history depend on whether the social scientist is located in, say, Europe or Africa, late 19th or late 20th century, and so on. What may have seemed an inevitable trajectory prior to the Bolshevik Revolution looks at best like a politically propelled idealization after the reversal and perversion of various socialist projects inspired by Marxism. This reflects the fact that Marxism is not simply an account of history but itself a part of history.

Emblematic of the ascendancy of emergentist over instantiationist thinking in our times is the decline of that pre-eminent closed science, physics, as the intellectual vanguard and financial leader of all the sciences. In the 19th and 20th centuries, physics had been the cornerstone of the instantiationist perspective, especially as the standard-bearer of *positivist* and *reductionist* ideologies (as discussed in the previous chapter), whereby disciplines would prove their progressive-ness by repeating salient stages in the development of theories and methods in physics.

It had helped that physics was traditionally a laboratory-based subject that not only sought laws for closed systems abstracted from locally variable effects, but also largely managed to insulate its own activities from their real world political consequences. The success of the US atomic bomb project in World War II is an example of the synergy resulting from this dual sense of autonomy. Indeed, it inspired Vannevar Bush's influential 1945 essay, *Science: The Endless Frontier*, which helped establish the US National Science Foundation. The dual autonomy of physics constitutes an "idealization" that has had both positive and negative import. Positively, the history of physics can be more easily told as a sequence of self-generated problems and solutions – the basis of Kuhn's paradigm-based theory of scientific change – than the history of, say, chemistry, biology, or the social sciences, where it is difficult to avoid the role of non-scientific influences on the research trajectory. Negatively, physics came to be overadapted to a state-protected funding environment that is gradually disappearing with the end of the Cold War.

In the "free market" of today's open science, it is much more persuasive to claim utility than autonomy, as illustrated in the race to map the human genome. The ability of physicists to demonstrate that just one more (and bigger) particle accelerator will answer age-old questions that only elites have been empowered to pose has been overtaken by biologists who claim that funding their research will enable ordinary people to customize their offspring. This shift from a vertically organized, theory-driven conception of science to one that is horizontally organized and driven by practical concerns marks science as undergoing a *secularization* comparable to the one which Western Christendom underwent, starting with the Protestant Reformation in the 16th century. The formal separation of Church and state initiated a period of religious evangelism, in which churches were forced to tailor the faith to fit their potential constituency, on whom they had to rely directly for material support (Fuller 2006a: ch. 5). So too with the post-physics, post-academic world of science.

As heir to the emergentist tradition, contemporary biology has exhibited a dual sense of "integration" that has made it "adaptive" in both a positive and a negative sense. On the one hand, biological research has increased our knowledge of the full adaptive capacities of humans and other species. On the other hand, biology's own research trajectory has been perhaps too adaptable to the interests of its host societies, which has resulted in a skewed knowledge base. Consider the Human Genome Project. Its large financial and cultural investment is predicated on its funders believing that difficult problems in social policy can be eventually solved – and maybe even pre-empted – by pre-natal genetic manipulation. Alteration of the physical environment or the interests of those who already populate it – say, in order to foster greater biological diversity – is presumed to be sufficiently complicated and expensive to be given secondary status in the research agenda.

In sum, the transition from closed to open sciences is epitomized by a major shift in the sense of "control over the environment" that is constitutive of scientific progress. From a preoccupation with predictive accuracy, scientists are now increasingly concerned with expanding the range of human adaptiveness to fundamentally unpredictable situations. In the physics-driven world of closed science, the main normative danger was that the artificiality of the laboratory would be used as a springboard for coercive social policies, as suggested by the phrase "social engineering." However, in the biology-driven world of today's open science, the main danger is the tendency to confer too much value on statistically normal behavior occurring in stable environments, so that robust survival ends up being amplified into some higher virtue like truth, goodness, and justice – sometimes called "sustainability." Part III of this book presents the explicit convergence – and perhaps sublation – of these two normative dangers through the increasing role of biotechnology in society's infrastructure. However, in the rest of this chapter, our attention will be focused on two rather different attempts to construct a political horizon for just this emerging situation.

2. Bruno Latour's *The Politics of Nature*: A Case of False Advertising?

Latour has gradually inched his way toward science policy in a series of books (Latour 1993, 1996, 1999). In the latest and probably consummate instalment of this project, *The Politics of Nature*, Latour (2004) reveals his membership in the somewhat touching but profoundly flawed tradition in French political thought that would cut the Gordian knot of historically entrenched oppression by establishing a new order in which the oppressed spontaneously embody the democratic ideals they have so eagerly longed to express. Latour stands out only in his belief that the league of the oppressed extends across species boundaries – and maybe even physico-chemical boundaries (e.g. from carbon-based organisms to silicon-based computers). It seems that to be eligible for the "parliament of things" that Latour would have preside over the "politics of nature," one will need to be sentient, with the understanding that the realm of sentient beings is likely to expand as we learn more about how things respond to their environments and hence become able to confer political significance on those responses.

Nevertheless, the mind boggles at exactly how deliberations in this ecologically correct assembly would look. How would claims to knowledge and power be transacted? How would matters be decided? How would consequences be judged? I shall address these questions first at the level of metaphysics and then politics. The former discussion draws directly on the previous section, in terms of which Latour's would-be politics of nature is an especially radical version of the nonstandard grand metaphysical strategy. The latter discussion will focus on the sense of inclusive representation promoted by Latour's "posthumanity" – or is it "inhumanity"?

2.1. What's so "natural" about The Politics of Nature?

In his earliest attempts to use the "nonstandard strategy" to extend political status to non-humans, Latour was widely read as trying to revive the Aristotelian doctrine of *hylomorphism*, the infusion of spiritual properties in material things, a mere metaphysical cut above primitive animism (Schaffer 1991). This conclusion was reached by interpreting Latour's move in terms of the standard strategy, whereby granting agency to objects such as computers and doors would increase the amount of agency in the world, thereby producing an image of the laboratory that Piagetian children in the first two stages of cognitive development might have, namely, a world populated by objects moved by ghosts. However, this absurd conclusion can be avoided by imagining – as the non-standard strategy would suggest – that there is a fixed amount of agency in the world, such that the more objects that possess it, the less of it each object possesses. And so, by spreading agency thin, the mystique of being an agent disappears (Fuller 1994). In a turn of phrase that would appeal to Latour's thermodynamically fixated philosophical mentor, Michel Serres (1982), the concept of agency is thereby "wasted": it can no longer be used as a principle of order that privileges certain objects at the expense of others. While the allusion here is clearly to entropy, the import of Latour's usage is to dissipate the structures of activity/passivity that are needed to maintain *any* power relations.

Another important precedent for this line of thought is Hegel's account of the progress of Reason dialectically unfolding itself in the world. As Hegel put it, originally only one person was free – the oriental despot – and everyone else was enslaved. Freedom under such circumstances corresponded to arbitrary will, as no one had the power to check the despot. An unconditional reasoner is thus able to behave irrationally – at least as seen from the standpoint of subsequent stages, in which this despotic concentration of power is distributed to increasingly larger numbers of people, until it is completely "democratized." In each stage, the character of reason and freedom is altered because one needs to act in the presence of others whose agency one had not recognized before. In Freudian terms, democracy is "socially sublimated" tyranny. But whether expressed as the spread of freedom or the dissipation of energy, this non-standard way of bringing about "the death of the subject" stands in interesting contrast to the more standard strategy of such recent French poststructuralist philosophers as Foucault and Derrida. They tend to presume that subjectivity will subsist unless explicitly eliminated or reduced to "effects" and "traces," while Latour follows Serres in supposing that a subtler but more effective way of erasing the significance of the subject is by flooding the world with subjectivity.

Latour (1988b) took most explicit aim at the standard metaphysical strategy by trying to subvert the classically privileged form of representation, *explanation*: more specifically, the image of explanation that has dominated Western science since the heyday of Newtonian mechanics, namely, an image that unifies by reducing, that explains the most phenomena by the fewest principles. This conception of explanation implies a radical asymmetry, whereby one representative (the explainer, or *explanans*) stands for as many representables (the explained, or *explanandum*) as possible. It is also the conception – the so-called covering law model – that is most familiar to contemporary philosophers of science who descend from the positivist tradition. To evaluate Latour's purported subversion here, we need to consider explaining as a social practice.

The point of explaining is to bring into relief properties common to a set of individuals that had been previously occluded. A typical result of providing an explanation is that these individuals are treated in a more uniform fashion than before, thereby reinforcing the appropriateness of the explanation to each new situation. This observation had been already made in 1960 by the speech-act theorist J. L. Austin, who was concerned with the "perlocutionary force" of a ramified linguistic practice like explaining. However, Latour advances beyond a speech-act analysis of explanation by revealing the amount of hard work that needs to be done in order for an explanation to succeed. In particular, the relevant individuals must exchange their voices for that of the explainer. This may be done with or without resistance from those individuals, but, in either case, the idea is that, as a result of the exchange, the individuals must go through the explainer in order to have their own interests served. At the very least, the individuals must find that the cost of contesting the explanation outweighs the benefits. In that sense, the explainer becomes what Latour (1987: ch. 6) calls an *obligatory passage point*.

The more ambitious the explanation, the wider variety of obstacles that the explainer is likely to encounter. For example, if you try to put forth an explanation that is common to objects that move on earth and in the heavens, you will probably have to contend with specialists on terrestrial and celestial motions, the legitimacy of whose careers may depend, in large part, on the two types of motions requiring separate expertises. You may thus need to make allowances to these specialists. Nevertheless, the successful explainer can keep these concessions hidden – in, say, *ceteris paribus* clauses that form the conditions under which the explanatory principle can be applied in particular cases. Ultimately, success in this arena is measured by the explainer's ability to get the locals to regard their respective clauses as *merely* local, in the sense of being relevant only for bringing to the surface principles that are already latently present in their locales. If the locals should combine to realize that what had been in their best interest not to oppose as individuals is now worth opposing as a group, then the explainer is faced with the amalgamation of anomalies that precipitates a Kuhnian "crisis." But, failing the emergence of class consciousness among the anomalous locals, their "tacit consent" enables the explainer to rewrite the history of her travails so as to make it seem as though disparate phenomena were spontaneously attracted to her explanatory principles, when in fact she had to negotiate with each one separately in their own terms. Latour speaks of this special power as the conversion of contact motion into *action at a distance*.

One suggestive characterization of this conversion is in terms of a Piagetian developmental transition from *spatio-temporal* to *logico-mathematical* cognitive orientations (Piaget 1976; cf. Fuller 1992). In the former orientation, the child treats objects as having a found order, or "natural place," and thus will maneuver around them so as not to cause a disturbance. In the latter, the child realizes that the objects can be moved from their original locations to a central place where the child can rearrange the objects as she see fits, which typically involves grouping some together and separating them from others. In this way, Piaget speculates, primitive set theoretic relations of inclusion and exclusion emerge, which are, of course, essential to any covering law model of explanation. And because the material character of the objects offer little resistance to the child's manipulative efforts, she can play with them without concern for the consequences of her interaction with them.

Piaget regards this sort of play as a primitive instance of pure theorizing, in that the child's manipulations are determined almost entirely by the rules of the game she has imposed on the field of objects. One obvious point that often goes unnoticed in discussions of Piaget's views, which nevertheless bears on our understanding of Latour, is that even when a child operates in the logico-mathematical mode, she continues to arrange objects in the world of space and time, but simply denies to herself that the actual arrangements she comes up with represent the full range of possible arrangements. By analogy, then, once an explainer has overcome an initial array of opposition, she will tend to presume maximum generality for the scope of her explanation, until she is forced to do otherwise. At that point, the explainer perhaps begins to take her own rhetorical reconstruction too seriously, believing that she can "act at a distance" without any "hysteresis" effects, that is, without any interference in the transmission of explanatory power from fundamental principles to disparate cases (cf. Georgescu-Roegen 1971: ch. 5). This superstitious phenomenon not only harks back to an earlier period in the history of science, but also remains common in contemporary historical accounts of science that speak of, say, the research trajectory of Newtonian mechanics as the application of Newton's principles to an ever wider domain of objects, as if the process were one of uninterrupted assimilation.

As already suggested, Latour's non-standard solution to the total-izing tendencies of explanation basically involves flooding the world with explanations, thereby eliminating whatever economic advantage a set of principles might have over a set of phenomena. Each phe-nomenon is to be given its own unique explanation, presumably one making plain all the effort it takes in that case to have the *explanans* represent the *explanandum*. The curious feature of this move is that it retains a vestige of the standard strategy's approach to explanation, namely, the explanatory principles still replace the things explained, though now on a one-to-one rather than a one-to-many basis. And so, while Latour's recommendation would remove the mystique from the practice of explaining, the fate awaiting the individuals on either side of the explanatory divide remains unclear. Will *explanans* and *explanandum* come to resemble one another, so that, as it were, puta-tive causes and effects will appear interchangeable? Apparently, Nietzsche countenanced this prospect in his original forays into deconstruction (cf. Culler 1982: ch. 1). Or will other practices emerge to reinstate the distinction that explanations draw? After all, nominalists from Peter Abelard to John Stuart Mill have believed that the homogenizing – and, in that sense, generalizing – tendencies of abstraction were inescapable features of the human mind. In that case, we might expect to find a mutant strain of explanatory prac-tice that appeals to the recurrent properties of "locality," "situated-ness," "contexture," and other tools of social construction in order to account for why things appear as they do: in a phrase, science as ethnomethodology.

What Latour seems to have retained from the standard metaphysi-cal strategy is the so-called correspondence relation, which ensures that nothing (important) is lost in the translation from *explanandum* to *explanans*, except, of course, the uniqueness of the *explanandum*, now that it has been reduced to yet another instance of the *explanans* in action. Latour thus envisages one-shot explainers eventually exhausting themselves by having to reproduce in their explanations that uniqueness as well. However, this conception of translation, which leaves no lasting impression as an activity (or "material prac-tice," as it is often phrased in STS) on the explainer and the explained, but only in their relationship, is little more than a conceit of logical positivism. In fact, the theory of translation with which practicing translators implicitly operate sets a more instructive example about how the non-standard strategy should proceed on matters of expla-nation. In particular, translators are aware that the consequence of any translation is to bring the two languages either a little closer together or a little farther apart. The former, an instance of *dynamic equivalence*, reproduces the pragmatics of the source language in the target language (typically at the expense of conceptual nuance), whereas the latter, an instance of *formal equivalence*, reproduces the semantic distinctions of the source language in the target language (typically at the expense of practical relevance). The difference that Bible translators draw between *hermeneutics* and *exegesis* intuitively captures what is at stake in the distinction: *texts for practitioners* vs *texts for scholars* (cf. Fuller 1988: 128–38). The point is that, as the process of translation is repeated between two languages, the languages themselves start to blend into each other in interesting ways, reflecting the balance of trade between dynamic and formal equivalents. I have spoken of this as an "interpenetrative" rhetoric that can bring forth comprehensive transnational trade languages such as Arabic and Swahili, not to mention the hybridization of existent global languages like Latin and English (Fuller and Collier 2004: ch. 2).

The Parisian collaborator on whom Latour relies most heavily for empirical confirmation of his metaphysical speculations is the engineer-economist Michel Callon. Together they have taken both Hegel and Darwin one step further by redistributing the voice of reason across not only class but also species boundaries – and not merely within the primate order. The interdisciplinary field of semiotics provides precedents for this semantic extension, as it routinely translates technologies from instruments of control to media of communication, or "sign systems" (e.g. Eco 1976). In a case study that continues to attract considerable controversy within STS (Collins and Yearley 1992), Callon (1986) argued that trawling technologies enabled scallops to articulate an agenda that was at odds with the agenda of the humans who would turn them into seafood. What would ordinarily be taken as a failure to control nature was thus rendered as a disagreement between two interlocutors, humans and scallops, who communicate in the language of "trawling" over boats and nets, much as two humans might communicate in English by telephone.

By so extending voices to scallops, Callon did not claim that we can communicate with the sea creatures as well as with other humans, but rather that our grounds for *intra*-species communication is really no better than our *inter*-species grounds. Callon was advising not to presume too *much* of our ordinary communicative skills. In both intra- and inter-species cases, the two interlocutors judge communicative success by their own criteria, which, in turn, depend on judgments about the degree to which the communicative medium is perceived as faithful to one's own wishes and the extent to which the interlocutor is perceived to have complied with those wishes. In other words, "scallop-talk" is important not for what it makes of our ability to communicate with scallops but for what it makes of the general criteria by which we identify the criteria of communication.

In effect, Callon converted the skeptical response to what philosophers call "the problem of other minds" into a methodological principle for sociology inquiry. If "communication" and "control" mark the two poles by which we distinguish our relations with humans and things, Callon purported to show that the poles are separated by a continuum with no clear dividing point. From the standpoint of political representation, it is worth observing that, on Callon's analysis, if the scallops do not display resistance, they are implicitly giving their consent to the trawlers, which is of course the default interpretation of political action in democratic regimes – for better or worse. In other words, Callon like most *de facto* democrats presumed what Jürgen Habermas (1971) would call a non-oppressive, or "ideal," speech situation, namely, one where if parties wanted to say something, they would freely do so. How well this generally characterizes the interaction between humans and non-humans (let alone between humans!) remains to be determined.

Nevertheless, a theory rather similar to Latour's is currently popular in primatology, which reaches its conclusions via Darwin, without any obvious help from Hegel or post-structuralism. A school of evolutionary psychologists based in St Andrews University, Scotland, have argued that a necessary condition for the development of animal intelligence is a network of social relations that is sufficiently complex to serve as a second-order selection environment – a "second nature," as it were (Byrne and Whiten 1987; Whiten and Byrne 1997). Thus, reason is domesticated, or "civilized" as it is exercised in the presence of others who are taken to be acting similarly. This is known as the "Machiavellian Intelligence" hypothesis, according to which the cognitive complexity of primates is a direct function of the complexity of their social organization. Such a spur to intellectual growth is aptly called "Machiavellian" because it requires second-guessing the response of one's fellows before one acts. Even if others are not exactly reduced to mere means for achieving one's own ends, they are neverthelesss always the medium through which those ends must be realized. To put it in Latourspeak: the smarter the primate, the more "obligatory passage points" she can pass before she is licensed to act.

Since the initial formulation of the Machiavellian Intelligence thesis, some primatologists have even begun to communicate openly in Latourspeak (e.g. Strum and Fedigan 2000), though the political precedent for this line of thought is traceable to the anarchist strain in late 19th-century Social Darwinism, including Herbert Spencer and Peter Kropotkin, both of whom defined evolutionary fitness in terms of the capacity for cooperative behavior: i.e. the service to others as a prerequisite to serving oneself. From that standpoint, those unwilling or unable to make an adequate contribution to the collective project (what Latour would call an "actor-network") are a drag on the entire ecosystem: they consume resources without producing anything in return.

2.2. What's so political about The Politics of Nature?

Turning more explicitly to the political side of Latour's science policy, the history of constitutional thought offers an important precedent. The late eighteenth century witnessed the rise of two paradigms of constitution-making, American and French, which drew opposite lessons from many of the same Enlightenment sources (Fuller 2000a: 131–2; cf. Elster 1993). The closed-door sessions of the Philadelphia constitutional convention of 1787 produced a document that doubted that spontaneous expressions of self-interest would serve the public interest without an elaborate system of checks and balances and sep-aration of powers. In contrast, the open door sessions of the Paris con-vention of 1789–91 produced a document that presented a vaguer, more sanguine and consensual sense of government operating in the public interest to realize "the rights of man."

Latour belongs to this latter tradition, whereby "representation" is conceived as "self-representation." And while it may play to the gallery – as it did in revolutionary France – to maximize self-representation, there is no guarantee that such representatives will do their constituen-cies, let alone the entire polity, the most good. After all, the oppressed are likely to be just as shortsighted as the elites, except that they lack the experience with wielding power against resistant parties that have honed the negotiating skills of the elites. In short, rather than worry about the resemblance between the representatives and the repre-sented, the political focus should be on the accountability of the repre-sentatives to the represented. This is the lesson the Americans early learned, which is why they have stuck to the same constitution (with periodic amendments), whereas the French have gone through a dozen in the last two centuries.

This last point bears on the fundamental flaw in Latour's project. Like his French forebears, who replaced their Constitution as they replaced their collective self-understanding, Latour believes that a change in the categories salient for making sense of social life *ipso facto* requires a radically new political order. This belief constitutes a much worse example of the "scientization of politics" than anything dreamed up by the neo-liberal followers of Friedrich Hayek (1952), who have been obsessed with the technologically enhanced authoritarianism that the social sciences seemed to license. Latour appears to have succumbed to the temptation of Hobbes: to have "representation" mean the same thing in both scientific and political contexts. I too have struggled with this temptation in my own work on social epistemology (Fuller 1988: 36–45). Needless to say, it is always easier to identify one's own errors in others.

To be sure, the scientific and political senses of representation do converge at one point: both are strategic abstractions from the welter of phenomena. However, there is the "temptation" to conclude that the point of abstraction in both cases is the same. It is not. Scientific representation is the reproduction of a source in some other medium, such as a trace of some event or a sample of some population, whereas political representation involves holding a medium accountable to a source. The one is about getting the origins right; the other about getting the consequences right. This helps to explain why evidence has an ambiguous role in politics. It is not because politicians are especially duplicitous or unrigorous but because they are more concerned with the future than the past – and evidence is always about something that has already happened, which of course may or may not be relevant to what happens later.

Francis Bacon (under whom Hobbes served as personal secretary) had tried to convert the temptation to conflate the scientific and political aims of representation into an overarching virtue by introducing the idea of "crucial experiment," which would allow the scientific politician to generate evidence specifically designed to test competing visions of the future – what scientists call "hypotheses" and politicians "policies." As it turns out, Latour (2004) deploys the expression "experimental metaphysics," but to refer to the need to track empirically the connections between the various interest groups in the parliament of things. The word "experimental" appears simply to alert the inquirer to remain open to the possibility of surprising interdependencies among otherwise ideologically opposed groups. Latour's intent here is clearly to facilitate the drafting of legislation that can accommodate the widest range of groups. However, it is not obvious that groups should be so easily co-opted into a future on the basis of little more than their having shared a common past. In other words,

Latour's "experimental metaphysics" may be "experimental" for STS investigators like himself but, with regard to the political agents under investigation, it is no more than a scientifically cloaked appeal to the path of least political resistance, if not conservatism more generally.

Perhaps the greatest advantage of the American over the French founding fathers was their ability to design a robust yet supple system of government capable of withstanding substantial, albeit unknowable, long-term changes in the Constitution of the polity. The secret to their success lay in having devised better procedures, not for representing society's various interest groups, but for discouraging those groups from becoming rigid in their conception of self-interest. The argot that has grown around American politics – including "pork barrel" and "logroll" – captures the incentives politicians always have to creatively rethink their constituencies' interests without resorting to revolution or civil war. Of course, a politician may cut one too many deals and effectively sell out his or her constituency's interests. But that is a matter of representation-as-accountability, the true business of politics, which is remedied by an electoral defeat rather than a redrafting of the Constitution to ensure better representation of the constituency.

One sign that Latour has not thought through clearly the radical implications of his politics of nature is his failure to make common cause with the one person who has actually – and often scarily – done the detailed work of marrying the political logic of representative democracy to a scientific understanding of those eligible for representation. I refer here to the philosopher of "animal liberation," Peter Singer, who is nowadays fashioning a "Darwinian Left" (cf. Singer 1975, 1999). Like Latour, only more concretely, Singer wants to extend representation to all sentient beings, most of which cannot participate in the conventional channels of political communication. Singer explicitly draws attention to two uncomfortable but incontrovertible facts: (1) neo-Darwinism does not support a clear distinction, let alone the privileging, of humans vis-à-vis other animals; and (2) the history of democratic politics has been marked by periodic innovations in the criteria for representation, for example the replacement of property ownership by "citizenship" requirements. Singer argues that together these facts imply that self-declared progressives should work toward accommodating animals to the political system just as earlier progressives worked toward accommodating workers and women (cf. Fuller 2006b: esp. chs 9–10).

"Accommodation," however, is a euphemism, since in a world of scarcity, Singer is clear that some humans – unwanted and disabled – will have to yield their rights in his brave new polity to healthy and wanted non-humans. If nothing else, Singer is to be credited with having the courage of his convictions, even though it leads him down the path of racial hygiene, for which he has been denounced in the German-speaking world as a Nazi (Jasanoff 2005: 168–73; cf. Fuller 2006b: ch. 14; Proctor 1988; Weikart 2005). In contrast, Latour appears satisfied with operating at a level of abstraction that generates a frisson in his readers without forcing them to face the hard political choices involved in designing a true parliament of things.

Is Latour simply naive or disingenuous? He begins:

> If I have no authority of my own, I nevertheless benefit from a particular advantage: I am interested in political production no more and no less than in scientific production. Or, rather, I *admire* politicians *as much as I admire* scientists. Think about it: this twofold respect is not so common. My absence of authority offers precisely the guarantee that I will not use science to subjugate politics, or politics to subjugate science. (Latour 2004: 6, italics original)

Consider, as a point of reference, that someone else who could have written Latour's words was Max Weber, the great sociologist who delivered two addresses toward the end of his life entitled "Science as a Vocation" and "Politics as a Vocation." Had Weber uttered Latour's words, he would have meant that science and politics are two separate but equal domains of social life that operate according to mutually countervailing norms: the scientist pursues the truth regardless of political consequences, whereas the politician pursues an objective even if it requires compromising or denying the truth. The former is principled like Galileo and the latter adaptive like Bismarck, both in the face of changing circumstances. Moreover, Weber's observations were made in a prescriptive, not descriptive, spirit: however much scientists and politicians ease into each other's turf, the fate of humanity rests on a managed conflict between those fixated on abstract ideals (i.e. scientists) and on concrete results (i.e. politicians). As one vocation colonizes the domain of the other, civilization is brought to the brink of chaos. Weber spoke in the aftermath of World War I, in which his native Germany, the principal aggressor (and ultimate loser), had been backed by what was then the world's premier scientific community.

However, Latour conceptualizes matters rather differently. His equal admiration of politicians and scientists is born of a belief that, at their best, they are doing much the same thing. It is important to stress that Latour is not talking about how politicians and scientists justify their respective activities, which are admittedly cast in divergent terms, but how they behave on the ground: Galileo may have felt that his experiments and observations provided the royal road to the truth, but he was happy to stretch his arguments beyond the limits of his evidence in order to score rhetorical points against opponents. Likewise, Bismarck may have been the ultimate "power-politician," but his was a pursuit of power in the service of a united Germany that he genuinely believed had a world-historic role to play. Thus, Galileo turns out to be more adaptive and Bismarck more principled than they appear in Weber's account. Latour speaks of this as the "hybrid" character of what passes in francophone circles as *écologie politique* (Whiteside 2002). His hybrid hero turns out to be Louis Pasteur, whose experiments triggered a revolution in the management of agriculture, industry, and public health at once more effective and more peaceful than could be imagined, let alone executed, by even the greatest of French politicians (Latour 1988a). What is frightening in all this is that Latour seems to admire scientists mainly for their ability to change the world substantially without anyone ever holding them accountable, as the change is never subject to a decision by those potentially – let alone actually – affected by it.

So, are we doomed to choose either the brute candor of Peter Singer's scientifically informed political decisions or the charming stealth of Bruno Latour's politically freighted scientific practices as alternative constitutions for the democratization of science? Singer's open-faced arguments at least exhibit a sense of intellectual responsibility not so evident in Latour's. However, another slant on the issue was revealed to me by the leading Canadian sociologist of science, Yves Gingras, back in 1992, when Latour's politics of nature received its first public hearing in English – and before anyone had thought of making the comparison with Singer's politics. Gingras observed that animal rights activists may simply mobilize animals as resources to consolidate a power base they would otherwise lack. In other words, the most perspicuous interpretation of their activities may be simply as another moment in the history of democracy that alters the political currency to enable a larger portion of *humanity* to participate, according to principles that confer an advantage on the newcomers. Such an interpretation fits well with my own "republican" approach to the governance of science, which puts politics in the business of expanding the circle of human representation by engaging in periodic, and often radical, revisions of the salient features of human engagement.

3. The Author's Civic Republican Alternative

In contrast to Latour, I am more interested in the *governance of science* than the politics of nature (Fuller 2000a). Instead of Latour's diffuse, metaphysically inspired notion of democracy, I propose one grounded in a specific vision of democracy drawn from the history of political thought and practice: *republicanism* (Pettit 1997). In practice, it amounts to a more explicit articulation of Karl Popper's "open society" (Jarvie 2001, 2003). The combination of individual freedom and collective responsibility that are the hallmark of the republican regime is epitomized by the slogan, "the right to be wrong," which I hold to be essential to the spirit of the scientific enterprise. The social system of science performs optimally when it not merely tolerates but thrives in an environment where its participants can err with impunity. Historically, republican regimes have worked only when its variously interested members have been of roughly equal political and economic status and equally threatened by an external foe.

The ideal republican regime is one of perpetually managed conflict, in which citizens, by contesting each other individually, are better able to contest their common enemy, be it defined as a foreign invader or, in the case of science, the ever-present threat of falsehood. In practice, this means that the regime aims to jointly realize two tendencies associated with the semantically rich phrase "being open-minded" that have historically cut against each other: on the one hand, to contest claims as vigorously as possible, and, on the other, to change one's mind as publicly as possible. Though rarely achieved either in politics or science, the joint realization of these two tendencies makes for the best in both realms. Moreover, I believe that the capacity for republicanism marks human beings most clearly from other living things. However, after surveying the political character of my republican approach to science in more detail, I shall argue that changes in the scale and scope of the scientific enterprise may be subverting these very republican capacities.

3.1. The quest for a reversible science policy regime

Throughout my work on social epistemology, I have assumed that it is both possible and desirable to construct forums for "knowledge policy," understood in the broadest sense to cover both educational and research matters in all the academic disciplines. It would enable an entire society, regardless of expertise, to decide on which resources should be allocated to which projects, on the basis of which accountability structures. In this spirit, I have advocated the institutionalization of consensus conferences, more about which below.

This vision has been predicated on the existence of nation-states or transnational institutions with the economic and political power to secure both the integrity and the efficacy of the knowledge-policy deliberations. One may have doubts about entities with such powers in these postmodern neo-liberal times. However, a deeper worry about my position is that, very unlike Latour's, my model of science is too discourse- or theory-driven and hence neglects the material level at which science is both conducted in labs and applied in everyday life (Radder 2000). Will the endless debates promoted by republicanism really address the core issues of scientific governance? The implication here, which I strongly oppose, is that the impact of science is felt much more in terms of *technology* than *ideology* (cf. Gouldner 1976).

Science-based technologies have radically altered the human condition, but ultimately what is most striking about the role of science in contemporary society is the ease with which we justify courses of actions by appealing to scientific knowledge without any clear understanding of either its content or its effects, let alone the conditions of its production. Often philosophers finesse this point by saying we have good reason to proceed in such ignorance because scientific knowledge is "reliable" (e.g. Goldman 1999). However, this appeal to reliability is itself an article of faith, since not only is there no published track record for most scientific knowledge claims but also there is no agreement on how such a record would be assembled. In the end, we simply proceed on the basis of persuasive narratives punctuated by anecdotes, as told by well-appointed experts. Therefore, given the often conflicting and self-serving nature of expert testimony, I place great store on the constitution of forums for scientific debate to manage these differences, very much on the model of "separation of powers" and "checks and balances" that characterizes the US Constitution.

Moreover, this is not only a philosophical project, as in Habermas's (1981) "ideal speech situation," but one with clear practical precedents. Beyond the republican traditions of constitution-making, there is also the *consensus conference* model of scientific deliberation (Fuller 2000a: ch. 6). Sometimes traveling under the name of "citizens' juries," these forums involve a cross-section of the ordinary uncommitted public in drafting policy guidelines for legislation governing a technoscientific issue of broad-based concern. Before drafting the guidelines, the delegates are exposed to spokespersons for various expertises and interest groups, whom they may cross-examine. In recent years, many consensus conferences have dealt with the implications of new biotechnology for health policy.

Contrary to its name, the goal of a consensus conference is *not* the establishment of a substantive policy consensus, but only on the parameters for subsequent deliberation and action, a task to be concluded by the duly elected legislative body. In the course of the deliberations, the participants learn to differentiate their own preferences from the public interest, while they are in the process of integrating the expertises and interests to which they were exposed. In effect, the participants are integrated into the knowledge production process at the same time they are integrating disparate forms of knowledge that neither the academics nor the policymakers had yet integrated successfully. The result is the simultaneous unification of knowledge at multiple levels of social reality.

While the discourse of the consensus conference is focused on a concrete task, the task itself is one that, as political circumstances change, could be revisited in the future. Such *reversibility* of decisions is crucial to collective learning. Moreover, unlike the ideal-speech situation, consensus conferences do not presume at the outset that participants can distinguish their personal interests from a larger collective good. Rather, that distinction appears as one of the outcomes of participation. If there is a drawback to consensus conferences, it is that they have indeed been treated as ideal-speech situations, academic demonstrations that indeed lay people can deliberate intelligently about highly controversial technical matters, but without their outcomes being integrated into the state policymaking process (except in Denmark).

The need for reversible decisions is crucial to understanding the sense in which I am concerned with the "material" dimension of science. STS tends to dwell on the materiality of science from, so to speak, the "inside," namely, the phenomenology of resistance faced by scientific practitioners (e.g. Pickering 1995; Rouse 2002; for a critique of this approach, see Fuller 2000b: 130–3, 314–17). In contrast, I hold that the sense of science's materiality that is relevant for science policy decisions requires adopting a standpoint *outside* a flourishing research tradition. It pertains to the overall *irreversibility* of the scientific enterprise that is associated with the phenomenon of "progress." This is what historians popularly call "forward momentum," which economists have theorized as "path-dependence" (Page 2006). The basic idea is that actions taken at one point in history constrain the possibilities for future change. This narrowing of alternative futures is then easily (albeit superstitiously) seen as movement toward a preor-dained goal, such as the ultimate theory of some domain of reality or the universal provision of wealth.

This mistaken attitude is abetted by what social psychologists call "adaptive preference formations," according to which we maintain cognitive equilibrium by regularly revising our wants so as to appear satisfied by what we are already doing or is likely to happen to us (Elster 1983). Ultimately this feeling of irreversibility can be explained as the result of investments in certain trajectories that, alongside yielding some fruits, have rendered the cost of changing course prohibitive (Fuller 1997: ch. 5). It is the latest incarnation of an intellectual lineage that starts with Aristotle's *dynamos* and passes through Aquinas' *materia* and Marx's *Kapital*. In each case, matter is understood as exhibiting a "mind of its own" by virtue of its having been informed by human labor. In this context, the republican approach to science policy is a call to "dematerialize" science by making its future less dependent on decisions and investments made in the past. It unites the range of policies proposed in Fuller (2000a). The four main ones are:

1. The principle of *fungibility*, which would make the inclusion of alternative disciplinary perspectives increasingly salient as grant sizes grow.
2. The detachment of theoretical approaches from the expensive methods with which they have been historically associated.
3. The extension of "affirmative action" legislation to break up the "cumulative advantage" that certain individuals and schools acquire as knowledge production sites (more about which below).
4. Perhaps most importantly, opposition to the curtailment of "free speech," simply based on the perceived costs to the status quo of taking an alternative point of view seriously.

Fungibility is the crucial concept here. The term is broadly used in economics for the capacity of different goods to satisfy the same func-tion, where "function" may mean labor, use, or exchange value. Typically advances in applied science and technology make goods fun-gible, and hence to some extent replaceable or expendable. Marx's cri-tique of capitalism can be largely understood as directed against human labor, understood as a capital good, being rendered fungible. Nevertheless, the increasing reach of fungibility comports with a sense of scientific progress consonant with the Enlightenment's original ideals. As Cassirer (1923) observed, the history of modern science can be told as the successive replacement of substance-talk by function-talk, as exemplified by the supplanting of Aristotle's verbal accounts of matter in motion by the algebraic notation developed by Descartes, Newton, and Leibniz.

To be sure, fungibility can be spun to opposed ideological effects. For example, fungibility's value to a civic republican science policy is its principle of accommodating the widest range of inquiries under a common budgetary framework through a kind of intellectual trade-off scheme (Fuller 2000a: ch. 8). However, in the neo-liberal regimes that most often characterize science policy today, fungibility is the principle by which competing research teams purport to provide the same knowledge products at ever lower costs to a public or private client, usually by relying on short-term contract labor and computer simulations (Mirowski 2004).

My specific focus on the question of science's reversibility evokes an issue that has preoccupied Popperian philosophy of social science for the past half-century, that is, since Popper (1957). In this context, one may distinguish two rather contradictory elements in Popper's approach to irreversibility. On the one hand, irreversibility is something to avoid, specifically forms of social engineering committed to policies that could not be reversed in light of harmful consequences. On the other hand, irreversibility is also a genuine and perhaps even defining feature of scientific progress. However, presupposed in this notion is that a truth once discovered – something that Popper held could not be planned – can be suppressed only by artificial (aka authoritarian) means. These two conceptions of irreversibility are reconcilable, but only once we alter how we empirically mark progress in science. Specifically, we need to recognize that both the benefits and the harms of science policy are likely to be asymmetrically distributed across society, absent some state-like entity to enforce a symmetrical distribution in the name of "welfare."

Perhaps Popper failed to appreciate the tension in his conception of irreversibility because he was partly influenced by the version of neo-classical political economy that was pursued in his Viennese days by Ludwig von Mises and his student Friedrich Hayek (Hacohen 2000: ch. 10; cf. Sassower 2006: ch. 2). Like them, he was enamored of an idealized "public good" conception of knowledge production, according to which power is involved only in blocking, not propelling, the flow of knowledge. In any case, Popper did not see that many, if not most, of the negative consequences of a socially engineered science policy would be registered in terms of *opportunity*, not real, costs. In the case of opportunity costs, the resulting harms are not borne by those who benefited from the original investments (say, in the form of a tax or penalty), but by those who did *not* so benefit (say, as a failure to develop one's potential).

While this last point may seem obvious, Popper nevertheless appeared to suppose that science policy impacts *equally* on the members of the republic of inquirers, because he embedded the republic in a democratic society whose respect for the rule of law would provide safeguards against legislation systematically advantaging or disadvantaging certain sectors of society (Jarvie 2001). On the contrary, I would argue that, precisely because science is presumed to be an optimally self-regulating sector of society, the distribution of benefits and harms is always asymmetrical and, if left unattended, will probably track already existing class, race, or gender differences. Thus, if the voices of those who are harmed are not as loud as of those who are benefited, then a policy measure might be easily registered empirically as "progress," since most notions of progress presuppose a strong distinction between those ahead of and behind "the curve." In Marxist politics, it divides the party vanguard from the proletariat; in marketing, it divides the early adopters from rank-and-file consumers.

Implied here is a critique of Robert Merton's (1977) principle of "cumulative advantage," which implies that the apparent spontaneity of intellectual transmission – the feature that has impressed Popper and others as signs of knowledge as a public good – may merely reflect that some people are especially well-placed to receive the messages being sent. These people, in turn, set the standard by which others are then able to receive the messages and register appropriate responses. Perhaps a useful way to see my position here is as an inversion of what Latour (1987) takes to be salient about "immutable mobiles," those knowledge artifacts produced in the scientific laboratory that manage to circulate in society at large. Whereas Latour stresses their power to transcend and refigure contexts of usage, I would attend to the differences between contexts that the mobiles do and do not reach in their quest for immutability.

3.2. Rhetoric as an evolutionary precondition for science

Crucial to the rise of republicanism as a moment in the history of humanity is the importance of *rhetoric* in enabling us, in the guise of "citizens," to rise above the rest of the animal kingdom. Rhetoric, an innovation of classical Athens, was probably the first social practice to appreciate fully the strategic significance of distinguishing personally held views from ones that are publicly displayed. Because of Plato's and Aristotle's famous criticisms of rhetoric's role in fostering the manipulation of human behavior, rhetoric rarely receives credit for developing cognitive virtue, specifically by forcing agents to clarify (at least in their own minds) their true ends from the means by which they are pursued. The means by which an agent comes to find an end compelling may be immaterial – indeed detrimental – to that end's pursuit in the presence of others whose cooperation is necessary for its realization. In that case, more efficient, rhetorically effective means must be sought to achieve the desired end. Such calculations provide the basis for the social construction of *normativity* (Fuller and Collier 2004).

A striking example of this point is the committed racist who advances his views by invoking statistical differences between the mean scores of whites and blacks on IQ tests. The origin of the racist's commitment may be unrelated to its scientific merit, but the public version of that commitment – with its appeal to statistics – is clearly designed to elicit support from more than those who happen to share his prejudices. Yet, by resorting to statistics, the racist opens himself to criticism from people who assent to his position *only* because it appears to have the backing of statistical evidence. Thus, were it subsequently shown that the IQ variance *within* racial categories exceeds the variance *between* them, these people would immediately abandon their support. Nevertheless, at least in principle, the racist would be free to mount his case once again, in light of new evidence that he thinks will appeal to a broad constituency.

The racist's personal beliefs belong to what Karl Popper and other philosophers of science have called the *context of discovery*, while the racist's search for a publicly acceptable theory belong to what they have called the *context of justification* (Fuller 2000a: ch. 6). Popper (1972) locates beliefs and theories, respectively, in what he calls *world two* and *world three*. The fundamental issue at stake here is captured by my favorite attempt to identify humanity's *je ne sais quoi*. This moment pertains to scientific inquiry, which Popper famously regarded as philosophical criticism conducted by other means:

> Scientists try to eliminate their false theories, they try to let them die in their stead. The believer – whether animal or man – perishes with his false beliefs. (Popper 1972: 122)

In claiming that our theories die in our stead, Popper presupposed a strong distinction between *beliefs* and *theories*. Beliefs are products of our unique psychological encounters with the world, which include the influence of family, schooling, and voluntary personal associations. In contrast, theories are publicly expressed versions of beliefs that need not incur any liability on the part of the believer, typically because in order to become public, theories must have been expressed in a language that does not make reference to the theorist's own origins.

The appeal to a strong belief/theory distinction as a mark of humanity has a venerable lineage that reaches back from Popper to Goethe, Vico, and ultimately Plato. Yet it has remained a minority voice in the history of Western philosophy, which has generally tended to locate the warrant of knowledge claims in the relationship between theories and the reality they purport to represent, rather than between theories and the theorists who purport to represent reality. Invariably the capacity for *deception* (of both others and self) figures prominently in this alternative tradition as a defining human characteristic, since in everyday life deception is the most common activity that presupposes a space between the private and the public representation of thought.

To be sure, deception has its own Darwinian roots in sub-human animals. Here I return to the Machiavellian Intelligence hypothesis I raised earlier in relation to Latour. Both epistemological and ethical issues are bound up together in this Machiavellian definition of humanity. Epistemologically, a norm is implied about the value of distinguishing between what one knows in the short and long terms. In the short term, a given theory may be deemed false because it does not fit with current evidence or exigencies. However, in the long term, a descendant of that theory may turn out to be sufficiently correct to set the standard by which subsequent research and practice are judged. Most importantly, our humanity lies in the possibility that *this successful theory can be put forward by someone who had previously failed.* Here we arrive at the ethical point: at their best humans can tolerate substantial opposition to the status quo, even when that opposition seems to be empirically ungrounded. (Perhaps this gives some credence to the Christian doctrine of "turn the other cheek".) Thus, rather than killing the deviant theorist, societies have sometimes funded research designed to increase his theory's empirical warrant – which, of course, may or may not be forthcoming.

I have used the term *disutilitarianism* to characterize the ability of a society to benefit from an error – or, to put it more neutrally, a deviation – of one of its members (Fuller and Collier 2004: 273). In the management literature, this process is often assimilated to "organizational learning," but the term "disutilitarianism" retains the ambivalent moral dimension of the process, one which may involve onlookers exploiting the consequences of the misfortunes of others. But the result is more than simply a decrease in the likelihood that the same mistake will happen in the future. That much can be explained in terms of ordinary organic adaptation to changes in the environment. Rather, the idea here is that human societies routinely provide incentives for individuals to pursue certain lines of thought and action without concern for their own welfare, typically by encouraging them to undergo the sort of extreme conditions represented by a test, trial, or match.

In the annals of philosophical anthropology, there are at least three different angles from which to gain a general understanding of the disutilitarian character of the human condition: *dirty hands* (Jean-Paul Sartre), *moral luck* (Bernard Williams), and *tertius gaudens* (Georg Simmel). "Dirty hands" suggests that any improvement in the collective requires the commission of local harms (this is the secular descendant of *theodicy*, the attempt to explain the presence of evil in a world created by a benevolent deity). "Moral luck" captures the arbitrariness with which extreme actions either change the constitution of the collective or themselves become a locus of collective censure: what makes one such act seem inspired and another criminal? Finally, "tertius gaudens" refers specifically to the third party who benefits from the misfortunes of others, often simply by virtue of having arrived on the scene after the afflicted parties.

On a more strictly sociological plane, we can grade types of institutions by how they foster the risk-seeking behavior necessary for disutilitarian outcomes. These institutions answer the question: how are individuals encouraged to discount their personal well-being? The "degree zero" type of institution is one that promotes self-sacrifice by encouraging individuals to identify completely with the group. Many examples can be found in the annals of religion and warfare. In the German idealist literature, where this conception of self-sacrifice has been most explicitly justified, an individual's life is tantamount to an experiment in the realization of collective potential. At the other extreme is the institution of games, which encourage people to excel in environments that simulate and highlight only a few features of the "real world" – that is to say, a field of play – with the understanding that any win or loss may be reversed in the next round. Here the player adopts a role that is sufficiently distinct from their ordinary identity that the player's performance in a game ideally has no impact on his or her status outside the field.

On this scheme, business and science are as hybrids that fall between the extremes of self-sacrifice and games. Business is a bit closer to the former extreme, and science to the latter. Together they capture the space that Popper wished to map as emblematic of humanity's collective intelligence. It is perhaps no accident that something called "speculation" figures prominently in both activities. On the one hand, financial speculators calculate the portion of their wealth they wish to risk in a given business venture. Their rationality rests on their ability to make investments now that will not interfere with their ability to make investments in the future. On the other hand, the scientist speculates (or "hypothesizes") theories that are well-developed in terms of some recognized methodology but perhaps do not correspond to the scientist's deep-seated beliefs. As in a game, you can walk away from a loss and return to play the next day; yet, as in self-sacrifice, the loss remains one from which others can materially gain if they are better placed to exploit your errors than you are.

The unique cognitive position of *Homo sapiens* in nature may rest more on the implications that a false theory has for *who* is subsequently eligible to propose a successor theory than for the actual content of the successor theory. Indeed, a good candidate for an "essence" to humanity is the phenomenon of the *second chance*: to wit, someone who presented a theory once deemed false can later present another theory that is generally regarded as true, or at least an improvement on the collective knowledge of the society of which the individual is a member.

Were other animals shown capable of drawing so clear a distinction between themselves (i.e. their beliefs) and their self-representations (i.e. their theories), they would be justified as full-fledged participants in the distribution of rights and obligations currently restricted to humans. *Homo sapiens* as a species is distinguished by its members' ability to survive the representations they make and, more importantly, to increase their collective strength by registering the failures of individual self-representations. In this respect, rhetoric and disutilitarianism are the twin cornerstones of our species identity.

A genealogy of the second-chance character of the human condition would trace the social practices that have been designed to drive a wedge between people's thoughts and actions, so that the enactment of one's thoughts allows for future action in spite of criticism and even failure. The classical basis of republicanism, namely, the requirement of property ownership for political participation might make a good starting point, since even those who failed to get their way in the Athenian forum were able, in principle at least, to return to their estates at night. Outspoken citizens did not have to worry about the fate of their family, livestock, or grounds simply because they espoused unpopular positions; indeed, the competence required for managing an estate was seen as providing them with license to espouse such positions.

But what about societies in which one's economic status is not so secure? How does one simulate the material grounds of autonomy traditionally provided by property ownership? In fact, republican polities like Athens were rarely as secure as they aspired to be. Once again, the history of rhetoric provides guidance on how the difference between the ideal and the real was met. A speaker could invoke illustrious ancestors as virtual witnesses to support claims that his audience would otherwise find outrageous; for, if the audience opposes the speaker, it would be effectively renouncing its own past. Logicians nowadays demonize this strategy as the *argumentum ad auctoritatem* fallacy, since this line of reasoning typically fails to establish the relevance of past authorities to present concerns. Nevertheless, the fallacy's adaptive function in rhetorically hostile environments should not be underestimated. Once again, the evolutionarily salient point is *not* that audiences always accepted the appeal to ancient authorities; rather it is that when they did not, the speaker escaped with his life, as discussion was deflected to the relevance of the dead ancestors to the case at hand.

As the primary medium of persuasion shifted from speech to writing, autonomy was increasingly secured by methodical forms of *ventriloquism*, whereby one's own voice would be "thrown" to agents that provided "independent" corroboration for what one wished to say. Scrutiny would then be subsequently focused on the agents, not oneself. A speculative history of this process would highlight the European Renaissance and Enlightenment for perfecting two genres of methodological ventriloquism: *forgery* (whereby others are invented to speak on one's behalf) and *satire* (whereby others are invented to speak against themselves, thereby rendering one's own case easier to make). In the transition from a literary to a more strictly scientific culture, with its stronger fact/fiction distinction, forgery metamorphosed into the *ethnographic report*, and satire into the *experimental test*. Although a proper history of this transition has yet to be written, a sophisticated computer simulation of the dynamics of knowledge production and distribution has been built on the assumption that methodological ventriloquism has played the sort of crucial role suggested by this account (Carley and Kaufer 1993).

3.3. Has scientific progress become evolutionarily maladaptive?

If Popper and his distinguished precursors are correct about the mark of our humanity, then seen from a Darwinian standpoint, human beings are inveterate *selection dodgers*. Accordingly, both nature and nurture are challenges to be overcome through collective intelligence, not non-negotiable, externally imposed standards on the sort of existence we can have. But science, traditionally our best vehicle for this transcendence, has undergone a pernicious development over the last hundred years. From its original role in enabling us to adapt better to the natural environment, science has itself now become a major part of the selection environment to which we need to adapt.

This reversal of evolutionary roles marks a curious step back from the general picture of human evolution that culminates in the slogan, "let our ideas die in our stead." For example, in Popper's depiction, knowledge in its most primitive form was tantamount to humanity's survival strategy, originally restricted to our brains and later embodied in artifacts. But in any case, knowledge was pursued only in aid of survival. However, once the human ecology stabilized, leisure was available (at least for some men) to turn knowledge itself into a full-time pursuit, which involved the refinement, criticism, and ultimately transcendence of existing dogmas and techniques. I would argue that Popper may have let the story end too soon, since he failed to emphasize that even the pursuit of knowledge has material prerequisites and residues that over time can significantly channel and perhaps even pervert the enterprise.

In particular, as I earlier observed, what has been in the past regarded as signs of "progress" – namely, the seemingly irreversible character of scientific growth – may turn out to be an adaptive preference, whereby we subconsciously revise our wants to make it appear that activities in which we already have made substantial material and psychic investments provide optimal satisfaction for those wants. In this way, modern science can be portrayed as providing us with all the goods, services, and solutions that humanity has always wanted. Consequently, there are no incentives – and often no resources – for pursuing alternatives that would divert us from the chosen path. Evolutionary biologists would call this *overadaptation*, which may become a prelude to extinction, if the environment changes drastically and unexpectedly.

In its pure form, Darwinism states that species are blind to the environments in which their offspring are selected. Put vividly, individuals can only hope that their expressed traits are adequate to the world in which they find themselves. In contrast, Darwin's historic competitor, Jean-Baptiste Lamarck, supposed that a species can "intentionally" adapt to a changing environment, at least by offspring incorporating in their germ plasm the experience of their parents. However, the upscaling of the scientific enterprise over the last 150 years has landed humanity in a predicament that may well be Darwin's revenge on Lamarck. On the surface, Lamarck was much too modest in thinking that a species can gain control only over the conditions of *its own* reproduction. After all, with the help of science, *Homo sapiens* has managed to exercise control (wisely or not) over the life conditions of a great many other species, not to mention the physical conditions of the planet more generally. But at a deeper level, we may have fallen short of Lamarck's criterion of species intelligence, since it is by no means clear that our cognitive grasp over *our own* life conditions is as secure as our grasp over those of other species.

This blindness may be seen as Darwin's revenge, with which sociologists and economists have traditionally tried to cope in terms of the "unintended" and "unanticipated" consequences of putatively intelligent human action. A most peculiar feature of the history of our species is that the most important changes to our living conditions were never meant as they turned out to be. Arguably the ease with which we attach the prefix "un" to "intended" and "anticipated" merely reveals the state of our denial of our animal lineage. While such an interpretation might be conducive to radical free marketers like Friedrich Hayek, I do not believe that what has been unintended or unanticipated must always remain so. There is evidence that humanity has collectively learned from past experience. To think otherwise, especially as a matter of principle, would amount to denying our capacity to rise above the other animals. Here an assessment of the history of technology in the transformation of our relationship to the planet is required (cf. Agassi 1985).

Our collective incapacity to anticipate long-term and large-scale change suggests an impoverished imagination for alternative future trajectories. This may be the direct result of overinvestment in certain research trajectories that effectively compel us to identify our own fate with their outcomes. Eventually we are blind-sided by overadaptation. To see science's overadaptiveness, we need to look at how research programs came to be constrained by the methods, techniques, and even equipment that first gave them empirical credibility. Economists call this process, by which contingencies in the past end up channeling future developments, "path dependence." For example, John Horgan (1996), a former editor at *Scientific American*, raised the simple but not unreasonable question of why the search for the ultimate unit of matter requires the building of ever larger particle accelerators, when other areas of science – including molecular genetics – have long made do with less costly computer simulations. After all, the particle accelerator is itself little more than an expensive simulator of the "big bang" from which all matter allegedly descends. The reason lies in the historic connection between particle accelerators and nuclear weaponry. Not surprisingly, with the end of the Cold War, large-scale physics research is discussed in terms of that ultimate symbol of overadaptiveness, the dinosaur.

Since the end of the Cold War, molecular biology has not only eclipsed high-energy physics as the favored Big Science, but has provided an alternative paradigm for the organization of the material conditions of scientific inquiry (Knorr-Cetina 1999). Befitting the current neo-liberal regime of science policy, instead of discrete state-funded research sites, the emphasis is now on geographically distributed projects that are jointly funded by the public and private sectors of several nations. But even though this change has engendered disputes over exactly who "owns" the scientific results (e.g. who can patent the genes or the drugs based on them), which simply did not arise under the old policy regime, that still has not diminished science's role in constituting humanity's selection environment. If anything, the biological turn has intensified that role so as to be profoundly threatening to the Popperian principle that our ideas should die in our stead.

Be it mere nostalgia or a going concern, this principle has traditionally enabled a clear choice in the political conditions for social change: one could try either to *persuade* the current population to change their behavior or *replace* them with other people who will spontaneously behave as they should, that is, *change minds or exchange bodies*. The former is usually seen as the more humane option because it engages in an explicit yet uncoercive fashion with the people who would be most directly affected by the change. Normal political argument is the paradigm case of this option, as it deals in a rhetoric that the audience is free to accept or reject. In contrast, our brave new world of *bioliberalism* would blur the distinction between persuasion and replacement (Fuller 2006b). Nevertheless, before proceeding further, it is worth recalling that philosophers of science have long discussed the replacement of people as a strategy for inducing radical conceptual change. Contrary to Popper's view about the openness of scientists to ideas that overturn their assumptions, Kuhn argued that new ideas usually require new people to carry them forward. Thus, according to Kuhn, an old paradigm only dies once its proponents are no longer around to obstruct the path of the new generation. Such a close identification of individuals with their ideas is anathema to the point of view advocated here, but it does bring scientific selection much closer to ordinary natural selection (Fuller 2003a: ch. 3).

If asked 50 years ago, probably I would have located the blurring of the difference between persuading and replacing people in brain-washing or electrochemical therapy. However, today it is possible to cite practices that attempt to improve upon Darwin's theory of evolution by natural selection. Suppose the nation-state – now armed with the results of the Human Genome Project – is included as one of the selection pressures on individual reproduction patterns. Should people be persuaded either to increase or curtail their reproductive capacity in order to improve the overall condition of humanity (or at least of those living in the particular nation-state)? This question already started to be asked – typically under the auspices of scientific socialism – by the end of the 19th century. In that context, the state would attempt to alter the rates at which different classes or races reproduce themselves, typically by providing attractive financial incentives to the target populations.

Do such efforts count as an innovative form of persuasion or a subtle form of coercion? Much depends on whether the targeted people are being made an offer they can, in principle, refuse. In my own republican terms: do they retain the right to be wrong? If refusal is not an option, then coercion is in effect. In the 20th century, this extension in the scope of political argument into the sphere of reproduction acquired a humane face from that unique marriage of the welfare state and eugenics known as the "Swedish way," which was promoted on the international scene by the husband-and-wife team of Gunnar and Alva Myrdal. For sociologists and social critics seeking a precedent for the "brave new world" of social policy on which we are embarked, the Swedish way is likely to prove much more illuminating than the more coercively eugenicist pretensions of, say, Nazi Germany or the Soviet Union. Here we enter the historically resonant discourses of *inheritance* (cf. Fuller 2000a: ch. 8).

The normative question that begs to be asked in this context is whether appealing to people's sense of economic vulnerability is an appropriate basis on which to secure consent for altering their (reproductive) behavior from the path it would have otherwise taken. This question should resonate with any social scientist who has ever had to provide incentives to potential research subjects for their research: is informed consent reducible to exchange relations?

There is a postmodernist answer to this question that receives some support from modern biology: namely, the premise of the question is wrong, since it is far from apparent what course of action people will take before they encounter the actual conditions of their environment. Manipulation is an ethical problem only in a world where people begin with clear lines of thought and action, and underhanded means are then introduced to alter them. But without this clarity at the outset, one's sphere for thought and action remains open-ended: the benevolent eugenicist's appeal is then simply one among many facing the social agent, who is ultimately responsible for deciding whether to construct his or her situation as one of "economic vulnerability." While this line of argument is normally invoked to cast away the bogey of "determinism" in both its social and biological guises, at the same time it invites the intervention of those able and willing to leverage existing power imbalances to their advantage, short of violating the physical integrity of the target population.

Traditionally, the greatest fear surrounding the biologization of social life has been that actions previously regarded as within the control of human agents will be assigned to factors which, if not strictly superhuman, are "infrahuman" and therefore beyond direct self-control. According to this dystopic vision, medicine would eventually replace law as the liberal profession entrusted with administering significant forms of deviance. However, as our knowledge of genetic mechanisms becomes both more comprehensive and more normalized, legal and medical modes of administering to deviance may come to be synthesized in new forms of self-surveillance that enable, say, the culpability of individuals to be judged according to their genetic make-up, which they are then expected to know (and hence hold in check) before they act. In that case, the specificity of one's biological knowledge would effectively become the measure of one's sense of responsibility (Fuller 2006b: ch. 5).

3.4. Reflections on the politics of the bio-social boundary

In earlier work (e.g. Fuller 1992, 1993), I have called myself a "sociological eliminativist," that is, someone who believes that fundamental ontology consists of exclusively sociological categories, in terms of which explanations of scientific activity should be cast. In light of the foregoing discussion, it is worth reviewing my reasons:

1. The traditional divisions in academic knowledge between the "arts" and "sciences," "natural" and "social" sciences, and so forth,

describe not a functional differentiation, but a mutual alienation of an original unified sense of scientific inquiry. In this respect, I see STS not as yet another discipline, but a vehicle for the reintegration of academic life.

2. As the natural sciences have come to be dominated by laboratory-based methods and, more recently, computer simulations, the sense of their access to something that might be still called "nature" has been attenuated to the point of becoming little more than a misleading metaphor.

3. When socially relevant properties are perceived as rooted in "nature," that perception too often becomes a license for politically extreme behavior, ranging from neglect, isolation, and extermination (when nothing can be done to change "nature"), to manipulation, coercion, and other interventions without consent (when something specific can be done to change "nature").

Point (3) is especially relevant to the current discussion. In the last half-century of debates over sociobiology, the racial component of intelligence and genetic engineering, political progressives have been sorely tempted to silence their less "politically correct" opponents by invoking the ideological character of biologically based policy arguments. However, the silencing strategy – whether it involves ignoring or preventing speech – is not only inherently demeaning but often unwittingly generates sympathy for the silenced parties (Noelle-Neumann 1981). My civic republican science policy would not only block this regrettable strategy, but also encourage the proliferation of unpopular and even offensive viewpoints. My proposal here has two provisos: first, that the critics of these viewpoints are provided equal license and encouragement; second, that any automatic links are severed between arguments about "nature" and policies that might be taken on the basis of those arguments.

Much of the conflict between sociology and biology on policy matters stems from the level at which they place the efficacy of policy intervention, which in turn reflects deeper ontological differences over where to locate "nature," understood as some ultimate barrier to directed change. To be sure, actual sociologists and biologists occupy intermediate positions, but by considering ideal types of the two fields, we can get a good sense of where each locates the burden of proof in policy arguments. In sociology, individuals are presumed to be tractable through education, persuasion, incentives, and coercion, while entire societies are presumed to be naturally occurring entities that may or may not reflect the interests of their constituent individuals. In contrast, biology presupposes that efficacious change can occur at the collective level, say, by regulating the reproduction patterns of individuals, but that the nature of the individuals concerned is largely fixed by their genetic make-up.

Because sociologists have tended to regard entire societies as the ground of social reality, they treat individuals as unstable and even transitional entities (say, between generations of societal reproduction). Again, in contrast, biologists have tended to regard such macro-level categories as "species," and certainly "society," as only more-or-less stable emergent products of the interactions of rather more durable individuals. However, the difference between these ideal types is a century-old atavism that may be best abandoned in the 21st century. Indeed, Michel Foucault (1970) was probably right when he argued that the need to circumscribe the "human sciences" from the rest of naturalistic inquiry is a 19th-century idiosyncrasy that started when Kant coined "anthropology" in 1795 and had already begun to unravel with Freud in 1900. In that respect, I endorse projects that call themselves "biosociology" or "sociobiology" (Fuller 2006b). At a policy level, the boundary between the two fields was first blurred when governments offered financial incentives to encourage or inhibit the production of offspring in the early 20th century, and now is being completed as it becomes technically possible to design offspring while still in the womb.

In today's intellectual climate, "biology" and "sociology" stand less for distinct fields of inquiry than alternative political attitudes about how today's individuals should be oriented to future ones. In this context, sociologists too often urge that today's individuals adopt an ostrich-like attitude toward the increasingly available biotechnology. Yet, historically, sociology has contributed to the empowerment of humanity by arguing that, given the political will, ascendant forms of social technology can be used to advance their interests. There is no reason to arrest this tendency now. Sometimes it is said that the latest developments in biotechnology encourage people to act as gods, but then the same was said about mass literacy in societies where the ability to read was restricted to clerics with exclusive access to the sacred scriptures.

The only real problem here is a rhetorical one, namely, that persuasive appeals to "nature" have tended unwarrantedly to license the suspension of democratic political processes. For example, even if we granted that Euro-Americans score consistently lower on IQ tests than Sino-Americans (as the Canadian psychologist Philippe Rushton has maintained), and those tests are taken to be adequate measures of socially significant forms of intelligence, we would still need to decide whether resources should be allocated to compensate for the deficiencies of the Euro-Americans or to enhance the strengths of the Sino-Americans. Here the usual political, economic, and philosophical problems of distributive justice come to the fore, and we only pervert the public character of science if we think that they can be circumvented by presenting some robust empirical findings. The typical response of scientifically informed progressives to such biologically based policy proposals is simply to challenge the very robustness of the findings by faulting the methods used to obtain and analyze them. However, as STS researchers know (as the principle of "symmetry"), this is a risky strategy because it opens the door for more "politically correct" research to be examined with the same degree of scrutiny. As we shall see in the next chapter, this lies behind much of the so-called "Republican War on Science" in the USA: what goes around comes around, so it would seem.

In light of the forgoing discussion, the following four items should feature high on the agenda of normative social and political theory in our emerging bioliberal era:

- an examination of the material conditions under which people are forced to make choices about their own lives and those of future generations;
- a condemnation of the practice of involving people's economic circumstances in such choices – people should always be made offers that they *can* refuse;
- the provision of general education in biology commensurate with the growing level of self-surveillance required by the law's conception of personal responsibility – people should not be held accountable for things (even about themselves) that they were not in a reasonable position to know;
- use of the locution "forced to choose" as a semantic stopgap against the libertarian rhetoric that is likely to be pervasive in these times.

In Part III of this book, we shall return to these issues, especially as they are affected by developments in biotechnology and the potential revival of a social engineering mentality.

4

The Politics of Science Journalism

1. Science Journalism's Critical Deficit

Most STS research is founded on the idea that science and politics are inextricably bound together, at least in today's world and probably throughout history. But does it follow that science criticism is reducible to political criticism – or vice versa, for that matter? The practice of science criticism as political criticism is familiar from the critiques normally lodged against Nazi and Soviet science. The suspect nature of the science cast aspersions on the legitimacy of these regimes. But the sting of such criticism is typically felt, like the Owl of Minerva's presence, much more strongly after the formal object of criticism has left the scene. It is now easy to forget that many aspects of Nazi and Soviet science were praised in their heyday – even by their ultimately triumphant opponents in the West. This suggests that the validity of the science and the politics are, at least to some extent, decided together. But what about when political criticism functions as the basis for science criticism, such that the revelation of political motives is used to discredit the science?

This practice is already inscribed in US law, especially after *Edwards* v. *Aguilard* (1987), in which the Supreme Court drew on a radical reading of constitutional separation of state and Church to rule that the mere presence of religious motives was sufficient to disqualify creationism as a theory fit for high school science classes. The actual scientific content of the course materials and associated teaching practices are deemed irrelevant, if they can be shown to have been religiously inspired. In effect, the Supreme Court licensed the commission of what logicians call the "genetic fallacy" (cf. Fuller 2003a: ch. 15). Nowhere has this sensibility been felt more strongly than among liberal journalists, who have engaged in a veritable witch hunt to track down the religious and otherwise nefarious right-wing political sources for opinions that go against the scientific orthodoxy, on matters ranging from the teaching of biology to research into carcinogens and climate change. And because those sources are rarely hidden, it is easy to use that information to undermine the credibility of the ideas and inferences being proposed. What should STS make of this profoundly "asymmetric" practice of demonstrating political and scientific entanglement? I argue that such asymmetry constitutes a *critical deficit* in science journalism.

"Asymmetry" in this case refers to the practice of holding heterodox right-wingers accountable to the standards of orthodox left-wingers – but not vice versa. A celebrated exemplar of this strategy is *Washington Post* reporter Chris Mooney, who has recently documented the last two decades of science politicization in the best-selling *The Republican War on Science* (Mooney 2005). His book has been warmly received by those on the American left whose suspicions it amply confirms. But is there something genuinely newsworthy in the fact that science policy turns out to be politics-as-usual conducted by other means? I believe that at best Mooney has proven a banality, and at worst he encourages the contrasting illusory prospect of a science policy shorn of party politics – but just so happens to conform to liberal prejudices. By preaching so eloquently to the choir of disenfranchised American liberals, Mooney may have done more harm than good, obscuring a much deeper "secularization" of scientific authority that transcends party political differences.

Mooney's concern feeds into what the astute US historian Richard Hofstadter (1965) called "the paranoid style in American politics." The style is tied to the founding idea of the US as making a fresh start in the history of politics, specifically so as not to repeat the mistakes of the past, which included the establishment of a state Church in the colonists' British homeland. American paranoia runs very deep, perhaps most viscerally in the anti-Catholicism that routinely surfaced in political campaigns until John Kennedy was elected as the first Catholic president in 1960 – even though Catholicism has been the religion with the largest number of adherents for most of American history. But it has been subsequently present in court rulings such as *Edwards* v. *Aguilard*.

Against this cultural-cum-legal backdrop, we must also recall that the Republican Party has dominated US politics at the national level since 1968. The only interruptions in its reign have occurred when the "LBJ Factor" has kicked in – that is, a president in the mold of Lyndon Johnson, a folksy liberal from the old Confederate South (Jimmy Carter in 1976 and Bill Clinton in 1992). But those fleeting episodes were overshadowed by the Republicans' success in dismantling the "welfare-warfare state" which the Democratic Party established in the aftermath of World War II, and which had reached its peak in 1968, while retaining the full force of the "military-industrial complex" as America's main political and economic driver. Along the way, the Republicans have "politicized" science, in the sense of systematically supporting science policies that foster their party's agenda, an uneasy but surprisingly durable coalition of political conservatives ("the religious right") and economic libertarians ("big business").

Most of *The Republican War on Science* is about self-identified armies of scientists and policymakers. This is very much of a piece with the elite nature of political struggles in modern complex democracies. Nevertheless, these elites are a fraction of all the people whose cooperation is necessary for any policy to take effect. Mooney's oversight, which admittedly is characteristic of most contemporary science journalism, would never happen in political journalism (cf. Fuller and Collier 2004: 192–4). Imagine a journalist covering an election who reported the opinions of candidates and party operatives, and then turned only to think-tanks for assessments of the merits of the party platforms: no public opinion polls to establish the breadth and depth of voter sympathies; no probing interviews about which campaign issues really matter to voters. The natural conclusion to reach is that such a journalist has allowed herself to be drawn into the vortices of the spin doctors, whose combined judgments may or may not bear some resemblance to the election outcome.

If authors were judged by the quality of insight expressed in their epigraphs, Mooney would be deemed profoundly naïve about how science works. His level of naivete about the sociology of science would be unbecoming for a journalist operating in any other field of inquiry. Readers of Mooney (2005) are initially regaled with an epigraph from Steven Pinker, the first sentence of which reads: "The success of science depends on an apparatus of democratic adjudication – anonymous peer review, open debate, the fact that a graduate student can criticize a tenured professor." The pages that follow clearly indicate that Mooney believes not merely that this is a normative ideal toward which science as a whole aspires or to which pieces of scientific research might be, in principle, held accountable. Neither point would be controversial. Unfortunately Mooney also seems to believe that this is how science *normally* works.

Journalists should be especially scrupulous about distinguishing what people do from what they say they do. The ethnographic methods so beloved in STS and the more qualitative reaches of social science are historically indebted to just such first-hand coverage of previously neglected features of the life circumstances of workers and immigrants in the late 19th and early 20th centuries. However, Mooney's trust in the peer review system is based purely on high-minded hearsay. So let me report briefly as an "insider" to the process, both in terms of my own personal experience as peer reviewer for numerous journals, publishers, and funding agencies and in terms of my historical and contemporary research into the topic, which has included a global cyberconference on peer review in the social sciences (Fuller 2002a: Appendix).

The only place a graduate student is likely to criticize a tenured professor – and live to fight another day – is an elite university, especially when the professor speaks outside his expertise (as Pinker often does). Moreover, this phenomenon bears no relation to the workings of the peer review process used to decide grants and publications. Contrary to the democratic image that talk of "peerage" connotes, relatively few contributors to any science are regularly involved in the process. For the most part, there are no conspiracies here. It is simply a pain to spend time evaluating someone else's work when you could be doing your own work. Peer reviewing is a mark of "good citizenship," a euphemism for sacrificing a bit of yourself for the collective enterprise to which all involved would contribute. There are rarely any formal incentives to participate in the process. Of course, the burden is eased for those who work in the same field – but then ethical issues arise (to be discussed in the next chapter): will you stymie your peer's publication so that you can be on record as having said something similar earlier? In any case, funding agencies and academic editors end up gravitating to a relatively small set of referees who exhibit both reliability and soundness of judgment. While this process may resemble capitalism's "invisible hand," it is hard to see how it would conform to any reasonable understanding of "democracy." It is surprising Mooney trusts Pinker as a source for the virtues of the peer review process, since Pinker's last four books, all best-sellers, have been with commercial publishers.

2. Rising Above the Science Policy Spin

Science journalists are more like philosophers of science than either probably care to admit. Both are involved in public relations work for science without pretending to be scientists themselves. Of course, journalists and philosophers differ in deadline pressures, but they are similar in structuring their narratives around events, ideally ones where a discovery precipitates a decision with momentous consequences for an entire line of inquiry. Who exactly makes the "discovery" is an interesting question, since it need not always be the scientists themselves. It could be the journalist or philosopher, who realizes that a specific moment marks a turning point in a heretofore open-ended situation. Much depends on how the event is framed: what exactly is "news" here? For example, what was newsworthy about the mapping of the human genome – that it was done at all or that it was the outcome of a race between a publicly and a privately motivated team, or perhaps that both teams "won" on their own terms?

That many – perhaps most – would regard the bare fact that the human genome was mapped as news indicates just how little the general public previously knew about how much scientists know about our genetic make-up. From a strictly scientific standpoint, if not the original mapping, certainly the subsequent sequencing of the human genome was little more than a mechanical application, charting territory whose general contours were already known. This has itself become a concern for the scientists who wonder whether molecular biology is simply playing the endgame of normal science (Gilbert 1991). That a public and a private team competed to map the genome speaks to the anticipated consequences for the biomedical sciences and biotechnology: there is potentially huge consumer value in the mapping, but who will pay for what to be done? Perhaps that is a more newsworthy item. But one might equally argue that the segmentation of the scientific reward system, whereby one team gets its intellectual property rights and the other its Nobel Prize, points to the deepest issue of all, one that threatens any unity of purpose that scientific inquiry might be thought to have (Davies 2001).

The question of intellectual integrity in both the journalistic and philosophical cases pertains to just how independent is your representation of science: are you doing something other than writing glorified press releases for thinly veiled clients? It must be possible to be pro-science without simply capitulating to the consensus of significant scientific opinion. With this in mind, I am struck by Mooney's professed journalistic method:

> Let me explain my principles for reporting on science. In my opinion, far too many journalists deem themselves qualified to make scientific pronouncements in controversial areas, and frequently in support of fringe positions. In contrast, I believe that journalists, when approaching scientific controversies, should use their judgment to evaluate the credibility of different sides and to discern where scientists think the weight of evidence lies, without presuming to critically evaluate the science on their own. (Mooney 2005: vii)

The rhetoric of this statement is a marvel to behold. Mooney begins by distancing himself from colleagues who think they can pronounce on scientific debates. So, it would seem, Mooney defers to scientists. However, his own stated policy is "to evaluate the credibility of different sides," which sounds a lot like constructing an independent standpoint from which to pronounce on scientific debates. Mooney may be arguing at cross-purposes, but I would applaud the latter purpose as befitting a journalist who aspired to be the Walter Lippmann of the science field. Unfortunately, in the same sentence, Mooney dashes this hope by cashing out his idea of "evaluation" in terms of simply reporting the considered opinion of scientists. Thus, to evaluate scientific merits of neo-Darwinism and IDT, Mooney's repeated practice is to ask neo-Darwinists their opinion of work by intelligent design theorists but not vice versa. The results should surprise no one. Such opinion may indeed be expert but it is unlikely to be unprejudiced.

Mooney takes the judgment of the National Academy of Sciences (NAS) as the gold standard of scientific authority in the United States. Yet, the NAS is nothing but a think-tank that Abraham Lincoln created to provide advice during the Civil War that has been increasingly called upon by various branches of the federal government to research and advise on science-based policy issues. It is a self-selecting and self-perpetuating body of advisers that is not accountable to rank-and-file scientists, let alone the electorate at large. If the NAS is supposed to function as a kind of legislature for the scientific community, then it governs without their consent. Indeed, like most science academies established since the Enlightenment, there is no mutual accountability between the NAS and the universities through which scientists are normally credentialed. This is not to deny that NAS members are typically very accomplished scientists. But it is not clear that the quality of a scientist's judgment is improved as her achievements are rewarded. On the contrary, both the rewarding community and the rewarded scientist may come to adopt a superstitious attitude toward everything the scientist thinks. The rewarders confer a halo effect on the rewarded, a compliment the rewarded return by mounting scientific hobby horses that threaten to distort science's overall research agenda.

A notable case in recent memory is the ill-fated Superconducting Supercollider, a Congressionally funded project to build the world's largest particle accelerator in Texas. It is conspicuously absent from *The Republican War on Science*, though its heyday occurred during the Republican presidency of the first George Bush. The NAS was strongly behind it, fronted by distinguished physicists like Steven Weinberg and George Smoot. The latter's work on cosmic background radiation (a key to understanding the aftermath of the Big Bang) was indebted to a satellite launched by NASA, another of the NAS's ongoing interests. This is clearly science done mainly by and for its elite practitioners, who then gesture to its larger "cultural value" to justify its support. Scientific elites, especially in physics, have adopted this bread-and-circus approach to rebrand the grounds on which they were given *carte blanche* in the Cold War era. As should now be clear in retrospect, the "Cold" of the Cold War referred to the intellect, rather than the body, as the terms with which the Americans engaged in conflict with the Soviets: larger particle accelerators demonstrated the nation's capacity to harness energy to deliver larger weapon payloads; longer space voyages demonstrated the nation's capacity to, if not outright colonize, survey extraterrestrial domains. In the postwar thaw, these deferred preparations for war against a foreign foe were redeployed for a more direct national conquest of the structure of reality itself. For scientists like Weinberg and Smoot, that was the whole point of the exercise all along.

There is no doubt that the Supercollider would have – and NASA has – produced good science. Indeed, good science can be produced about infinitely many things but budgets are limited and hence priorities needed. A science journalist should be sufficiently alive to this point to report consistently the likely beneficiaries and opportunity costs of alternative science funding streams. Much too often, Mooney writes as if the entire scientific community would benefit from one funding stream, while only pseudoscientists and their political mouthpieces would benefit from another. Then those falling into the latter category are formally identified and, where possible, the patronage

trail is charted. Were Mooney more sensitive to the institutionalization of science policy, he would have recognized the asymmetry of his
practice. More specifically, he would have realized that two federal
science policy bodies he holds in high esteem – the NAS and the erst-
while research arm of the US Congress, the Office of Technology
Assessment (OTA) – operated under quite different principles, which
came to the fore in the debates that eventuated in the termination of
the Supercollider.

The OTA, the agency more clearly influenced by social scientists,
tended to frame analyses of the science policy environment in terms
of a comprehensive statistical representation of the range of constituencies relevant to the policy issue: that is, including not only elite
but also more ordinary scientists. On that basis, the OTA suggested
that if the research interests of all practicing physicists are counted
equally, then the Supercollider should not rank in the top tier of
science funding priorities because relatively few physicists would
actually benefit from it (OTA 1991). I say "suggested" because,
whereas the NAS typically offers pointed advice as might be expected
of a special interest group, the OTA typically laid out various courses
of action with their anticipated consequences.

My guess is that Mooney fails to mention this chapter in the OTA's
short but glorious history because it helped to trigger the ongoing
Science Wars, which – at least in Steven Weinberg's mind – is led by
science's "cultural adversaries" in STS (Weinberg 1992, 2001).
Indeed, at least one prominent member of the field, Daryl Chubin,
was a senior analyst at OTA when its reports contributed to the
demise of the overspending Supercollider. Although Mooney is right
that both the NAS and OTA have often found themselves on the
losing side in the war for influence in Washington science policy over
the past quarter-century, their *modus operandi* are radically different.
According to the NAS, science is run as an oligarchy of elite practitioners who dictate to the rest; according to the OTA, it is run as a
democracy of everyone employed in scientific research.

3. The Real Problem: The Politicization or the Secularization of Science?

Even if Republican politicians have tried to commandeer the scientific
agenda for their own ends, there are two countervailing considerations
that stand against Mooney's witch-hunt journalism. First, like it or not,
politicians and not scientists are the chosen representatives of the

people. And, at least in the US, the ballot box more reliably removes sub-optimal politicians than peer review identifies sub-optimal science. Second, even the most competent scientists have rarely agreed on policy direction. While this may be unfortunate, to believe otherwise is simply wishful thinking born of nostalgia for Cold War science policy.

The first point requires some elucidation. Whether politicians are doing good or ill, they are checked on a periodic basis through elections. In contrast, once a publication passes peer review, it has to be actively purged from the scientific literature – typically by scientists in related areas who are motivated to find fault in their colleagues' work. If we conducted politics this way, the result would be vigilantism, as in countries with weak democratic procedures. This disanalogy between politics and science matters more now than in the past, when control over the resources needed to promote research programs was not so concentrated. Back then (perhaps as late as World War II), it was easier for academic outsiders – not least early supporters of Darwin among the "captains of industry" – to advance their scientific agendas.

In any case, politicians are accountable to specific constituencies in a way scientists, especially elite ones, never are. Politicians are ultimately in the business of promoting the public interest, and everything – including science – is a means to that end. Whether she decides to listen to the NAS or scientists aligned with industry lobbyists, a politician's fate is sealed in the ballot box of the next election. If a great many politicians who spurn the NAS win re-election, then the problem would seem to lie with the disgruntled scientists rather than the politicians: perhaps voters are happy to take risks that scientists find unacceptable. Indeed, perhaps voters are happy to remain ignorant about the exact risks because of goods that can be plausibly delivered in the short term.

Suppose either or both of these speculations is correct. Does this demonstrate the irrationality of the American public? Mooney himself prefers to point to the ignorance and duplicity of politicians, as if the citizenry, "properly" informed, would reach conclusions that coincide with those of the NAS (Kitcher 2001 indulges this fantasy under the rubric of "well-ordered science"). Either Mooney is being incredibly polite here or he simply has failed to think through the implications of his argument. Why does he not argue that a body like the NAS should function as a second Supreme Court, with the right of judicial review over federal legislation? After all, if US policymaking is really drowning in so much bad science, then would it not make sense to suspend some democratic control over the research (and teaching) agenda? In Mooney's depiction, the pervasiveness of the problem certainly rivals that which brought a cabinet-level Department of Homeland Security into existence!

My own heretical view of this situation is that even if US policymakers are influenced by a degraded form of science policy, it may matter much less than Mooney thinks because the checks and balances of the political system ensure that the potentially worst effects of such policy – just like the potentially best effects of excellent science policy – are attenuated in its many stages of implementation and administration. And if this is not enough, there is always the ballot box as the site of revenge on politicians who too closely aligned themselves with a failed science policy.

A historical reality check is useful here. Like so many others who fret over the current state of science, Mooney compares the Republican politicization of science with Lysenkoism, the doomed Soviet agricultural policy based on a version of neo-Lamarckian genetics that comported with the ideology of dialectical materialism but not with the facts of heredity. And like so many others before him, Mooney makes the mistake of concluding that the main problem with Lysenkoism was that it tailored science to fit a preconceived political agenda rather than allow science to speak truth to power. However, this conclusion only makes sense with 20/20 hindsight, since Lysenko and his Stalinist admirers were involved in at least as much self-deception as deception (Roll-Hansen 2005). Nevertheless, what could have been noted even at the outset – and had been noted by consistent opponents like Michael Polanyi – was that the Soviet science system did not permit the fair testing of Lysenkoist knowledge claims (Fuller 2000b: 152).

It is disingenuous to think that science policies will not have elective affinities with the interests of the dominant political party. Mooney admits as much in his close association of what he regards as good science with the interests of Democrats and moderate Republicans currently out of favor in Washington. The real question is whether a science policy, regardless of its political origins, is subjected to sufficient scrutiny on the path to mass realization. While it would be nice to require every policy to satisfy state-of-the-art tests before it is unleashed on the public, something comparable may be simulated by having the policy pass through many different sets of eyes (of, say, bureaucrats), each attuned to different interests and hence motivated to troubleshoot for different problems. And if real problems pass unnoticed, then there is always the ballot box – hopefully enhanced by the spadework of investigative science journalists!

In short, the lesson of Lysenkoism is to beware *not* the politicization of science, but the *authoritarian* politicization of science. The *democratic* politicization of science – of precisely the sort encouraged by the federalist construction of the US Constitution – is fine. This is not to counsel a panglossian complacency toward US science policy. But as it stands, the best course of action for those interested in improving the quality of science in policymaking is simply to try harder within the existing channels – in particular, to cultivate constituencies explicitly and not to rely on some mythical self-certifying sense of the moral or epistemic high ground. Sometimes it seems that the US scientific establishment and the Democratic Party are united in death's embrace in their failure to grasp this elementary lesson in practical politics.

This raises the second countervailing consideration: science, depending on how you look at it, is a many-splendored thing or a house divided against itself. It is not by accident that the NAS was formed during the Civil War. Warfare, in both its preparation and execution, has provided the only reliable pretext for consolidating national scientific resources, where scientists have arguably spoken in one voice. Otherwise, scientists have been loath to form representative bodies that go beyond narrow disciplinary interests, and these typically more at a national than an international level. Considering that scientific fields of inquiry have universalist aspirations, this sociological fact is striking – as well as having been an endless source of disappointment for J. D. Bernal and other Marxists who hoped that scientists could be organized worldwide to lead a proletarian revolution in the 20th century.

Indeed, Mooney's jeremiad against the influence of scientists in the pockets of industry might best be read as evidence that scientific competence is itself no guarantee of political allegiance. This is less because scientists compromise the integrity of their expertise than their expertise is genuinely open to multiple applications and extrapolations, which may contradict each other. Whatever "value-freedom" science enjoys lies precisely here. It arises as a by-product of the controlled settings in which scientific expertise is typically honed and tested. These always differ sufficiently from policy settings to allow for substantial disagreements. I would even say that much of what passes for "data massaging," whereby empirical results are revised to justify a preferred policy option, may be explained this way. The primary sin in this case is one of omission – namely, of alternative trajectories that may be plotted from the same data, which in turn forecloses the opportunity for serious criticism of the preferred policy.

The controversy over Bjørn Lomborg's *The Sceptical Environmentalist* (not mentioned by Mooney but discussed in the next chapter) provides an object lesson in this point for the ongoing debate over global climate change.

Mooney does not take seriously the idea that scientists whose research promotes the interests of the tobacco, chemical, pharmaceutical, or biotech industries may be at least as technically competent and true to themselves as members of the NAS or left-leaning academic scientists in cognate fields. Where these two groups differ is over what they take to be the ends of science: *what is knowledge for – and given those ends, how might they best be advanced?* What Mooney often decries as "misuse" and "abuse" of science amounts to his registering opposition to the value system in which many politicians and scientists embed scientific expertise. For example, a quick-and-dirty way to sum up the difference between scientists aligned with industrial and environmental interests is that the former are driven by *solving* and the latter by *preventing* problems. The former cling to what the transhumanist theorist Max More (2005) calls the *proactionary principle*, and the latter to the more familiar *precautionary principle*.

Industry scientists function against the backdrop of an endless growth economy in which the maxim, "Necessity is the mother of invention," is a source of inspiration not desperation: any new product is bound to generate new problems, but those are merely opportunities for the exercise of human ingenuity – not to mention the generation of more corporate profits. That certain people are hurt by such reckless innovation must be weighed against others who would have been hurt without it, as well as the likely costs incurred by the available policy alternatives. On balance, then, one needs to be proactive to avoid the costs of inaction. In contrast, environmental scientists presuppose a steady-state economy, where the ultimate concern is that our actions reflect a level of restraint compatible with maintaining a "balance" in nature. This vision tends to privilege our current understanding of the future, including future harms, even though in the long term our understanding is itself likely to change, as we learn more. Thus, there is a risk when going down the precautionary route that the only "steady-state" being promoted is that of our knowledge, not of reality itself, as we prevent ourselves from taking risks that might serve to expand our capacity for action. Of course, environmentalists rightly ask who has licensed industrial scientists to risk other people's lives in this fashion, and toward ends that after all guarantee profits only for their paymasters rather than progress for all. However, these very same critics typically would also curtail experimentation on animals for similarly risky purposes. The result looks like a fear-based policy of epistemic ossification that rivals the sort of "faith-based" science policy that Mooney decries in scientific creationists and intelligent design theorists.

I do not intend to resolve this conflict in scientific world-views here. Each lays legitimate claim to advancing both science and the public interest. To be sure, the priorities of each are different, especially with respect to intertemporal issues: i.e. the relation of the short-term and the long-term. Neither world-view is especially prone to malice or incompetence, but there are clear reasons why certain constituencies might prefer one rather than the other. Moreover, the end of the Cold War has made the need for choice more evident. From my inaugural professorial lecture in 1995, I have argued that the status of science in society is shifting from that of *secularizer* to that of *secularized*: the ultimate moment of sociological reflexivity (Fuller 1997: ch. 4, 2000a: ch. 6, 2006a: ch. 5). The basic idea is that without a state-backed unity of purpose for science, instantiated in a centralized peer-reviewed system of research funding, science is bound to gravitate in many different directions, according to the strength of competing constituencies. This is the pattern exhibited by Christianity, once the secular rulers of Europe no longer required the approval of the Roman Catholic Church. Many rival Christian churches emerged in this religious free zone, each directly appealing to people's interests, forgoing abstract arguments that in the past only served to exercise authority over those people. In such a market environment, the religious concomitant of secularization has been evangelism.

An analogous "science evangelism" is readily seen today in the eclipse of state-based physics-oriented research funding by client-driven biomedical research. Whereas the citizenry used to dispose of their taxes to fund science as insurance against the vague but real sense of nuclear annihilation, nowadays they conceive of science as a high-tech service customized to their wants and needs. Perhaps politicians and the general public seem so much less informed about science than ever before because decisions about science are being placed more squarely in their hands. This is similar to what happened once the Bible was translated into the vulgar European languages, and believers were empowered to interpret the text for themselves. In the past, one could simply trust a state-licensed, professionally sanctioned peer review system to apply good science in a good way to good ends. People may have been just as ignorant, if not more so, but it didn't matter because they never had to take the funding decisions themselves. Like a nostalgic Catholic who in the wake of the Protestant Reformation thinks Christendom can be healed by returning to the papal fold, Mooney would have us return to the science-authoritarian days of the Cold War, which was actually an aberration in the political history of science.

The Discovery Institute in Seattle, Washington, is one of many think-tanks trying to jump start the future of science for political advantage. Its concerns range from intelligent design theory ("Creationism 2.0" to its detractors), through the ethics of biotechnology, to the more futuristic visions of "transhumanism" associated with the artificial intelligence pioneer, Ray Kurzweil – sometimes all combined together (e.g. Richards 2002). While the Discovery Institute is normally seen as right-wing, its perspective cuts across conventional ideological lines, manifesting the curiously durable coalition in the US Republican Party (matched somewhat by the UK Conservative Party) of the forces of tradition and liberty that comes into its own in defense of freedom of conscience associated with religious expression (Fuller 2007).

In an era supposedly marked by the imminent convergence of nano-, bio- and info- technology research (e.g. Roco and Bainbridge 2002) that will demand a redefinition of the human condition, think-tanks like the Discovery Institute may steal a march on the NAS. This is not because the NAS does not uphold good science – of course, it does, measured by traditional disciplinary standards. Nevertheless, such an elite institution is unlikely to have its ear sufficiently close to the ground to know what is and is not feasible in the foreseeable future, which is essential for framing any general political guidelines for research support. (That the NAS does not move very fast is symptomatic. Generally speaking, the peer review system has served to stagger publication, so as to allow a critical mass of relevant researchers to become "pre-acquainted" with impending research findings. But as time-to-publication shrinks in even the peer-reviewed sectors of the Internet, the advantage accrued to those "in the know" shrinks.) In short, the best work may be currently done *not* by the "best people" at the "best places"! I do not necessarily celebrate the diffuse and largely unmonitored – and certainly unregulated – nature of emergent technoscientific trends, but it is clear that the leftists with whom I identify are unlikely to win Mooney's "Republican War on Science" by clinging to a nostalgic view of the authoritativeness of the self-selecting college of scientific cardinals represented by the NAS.

4. Conclusion: Is Science's Secularization Reversible?

In the annals of US science policy, the genius of MIT's Vice-President Vannevar Bush's *The Endless Frontier* lay in persuading policymakers in the wake of World War II that the surest route to produce science in the public interest is to let scientists decide the research agenda for themselves. Not surprisingly, he made the argument turn on national security, based on the distinguished academic scientists amassed at Los Alamos who built the atomic bomb. However, an alternative framework for federal science policy had been floated even before America's entry into World War II, by West Virginia Senator Harley Kilgore (Fuller 2000b: ch. 3). He imagined a "National Science Foundation" as an extension of Franklin Roosevelt's New Deal. Kilgore proposed a science board in which two scientific representatives would serve alongside a representative each of labor, agriculture, industry, and consumer groups.

Like most astute observers at the time, Kilgore realized that innovative scientific research in the US was being conducted off campus, as academics saddled with heavy discipline-based teaching loads were lured to informally structured interdisciplinary research parks like Bell Laboratories. He believed, I think rightly, that scientists – like other high-skilled workers – would naturally gravitate to the best labor conditions, which could eventuate in the evacuation of scientists from the public sector. Not only would it be difficult to monitor or regulate their activity, it would prove difficult to reap the benefits implied by the Constitution-enshrined idea of science as a "public good." Using the Great Depression that ended the post-World War I economic boom as his benchmark, Kilgore believed that without state intervention, science would simply exacerbate class differences in American society. So, one of his many science funding schemes involved treating science education as a form of national service, whereby the government would finance the training of academically suitable students on the condition that they would spend some years developing one of America's economic backwaters.

Kilgore is relevant because he quite explicitly wanted to politicize science – indeed, to mount an offensive against scientists' spontaneous free-market politics. Moreover, Mooney would have probably found Kilgore's politics attractive (I certainly do). Yet Kilgore had no doubt that good science could be done under both private and public regimes. However, by the time the vote on the establishment of the National Science Foundation reached the floor of Congress in 1950, Kilgore's proposal had come to be seen through Cold War lenses as "politicizing science" in a sense by then associated with Hitler and Stalin. In Bush's victorious alternative, the federal government created a protected internal market for scientific research and later (in response to Sputnik) education. This has proved very costly and, not surprisingly, with the end of the Cold War, the federal government has gradually allowed science to revert to the pre-war free-market state that Kilgore decried. Anyone interested in promoting good science in the public interest, then, needs to articulate a robust conception of the "public interest." The New Deal was the last time that occurred in the US outside a context of military preparedness. The legacy of that formulation is what remains of the American welfare state.

If matters are to improve, the scientific community must engage in better public relations. Unfortunately, this is easier said than done. Introducing the 2006 paperback edition of *The Republican War on Science*, Mooney makes precisely this point but then undercuts it by addressing his readers as the "reality-based community," a self-congratulatory phrase adopted by US liberals, in equal measures obnoxious and desperate – especially given that members of this community find themselves on the outside looking into the seats of real-world power. It is worth recalling that this phrase, perhaps with a nod to the psychologist Paul Watzlawick (who, with Gregory Bateson, invented the "double bind" diagnosis of schizophrenia), originated in a 17 October 2004 *New York Times Magazine* piece by Ron Suskind that included the following comment attributed to an aide of George W. Bush:

> The aide said that guys like me were "in what we call the reality-based community," which he defined as people who "believe that solutions emerge from your judicious study of discernible reality." . . ."That's not the way the world really works anymore," he continued. "We're an empire now, and when we act, we create our own reality. And while you're studying that reality – judiciously, as you will – we'll act again, creating other new realities, which you can study too, and that's how things will sort out. We're history's actors . . .and you, all of you, will be left to just study what we do."

In other words, the reality-based community consists of history's losers, the sort of people whose political imagination is likely to be fueled by that sublimated form of resentment Nietzsche called *Ressentiment*. What makes *Ressentiment* trickier than ordinary resentment is that those under its spell so fully accept their own inferiority that they end up treating it as a source of strength. For example, many of Mooney's fellow reality-based communitarians take the minority status of their common belief in evolution by natural selection to imply that the theory is much too sophisticated for ordinary folks to grasp; hence the need for better public relations. (Of course, this does not explain why the US, the world's undisputed superpower in science, is one of the few countries where evolution is *not* uncritically accepted.) Mooney himself is a bit more circumspect: he seems to think that people are not too dumb but rather too *busy* to worry about science policy. This then begs the question: why *should* people worry if more than one account of the origins of life is taught in high school science classes?

5

So-called Research Ethics

1. The Two Dimensions of Knowledge Policy

My own approach to STS has been organized around the concept of *social epistemology* (Fuller 1988). Social epistemology is epitomized by a pair of questions: *Is knowledge well governed? How would we know?* The main difficulty in addressing these questions is *not* how to identify the best regime for "science," understood in broad (Germanic *wissenschaftlich*) terms as the organized search for knowledge. The norms governing such a regime enjoy a wide consensus, from which I myself do not dissent (Fuller 2000a). The regime is a kind of a civic republicanism, what Popper called the "open society," whereby scientists constitute the universe of mutually accountable individuals pursuing divergent means in support of a common end, in which the "common weal" is replaced by "truth." (These individuals may be doing other things as well, but they must be doing at least this in order to comprise a republic of science.) But agreement on the normatively preferred regime leaves open two other matters: *the ultimate ends served by such a regime* and *the degree to which current scientific practices approximate the best regime*. The former concerns the *external*; the latter the *internal*, social epistemology of science.

External social epistemology treats science as a sub-system of the larger social system, while internal social epistemology considers science as a system in its own right. Disagreements in external social epistemology tend to presume a common understanding of the current state of science but differ over the ultimate ends that science should serve, and hence how the present should be projected into the future. In contrast, disagreements in internal social epistemology typically presuppose consensus over the ends of science, but the parties differ over the exact empirical state of the field. In legal terms, external social epistemology is a *legislative* matter, and internal social epistemology a *judicial* matter. I shall take the two matters in this order, but since many specific issues concerning legislative knowledge policy have been already addressed in chapter 3, I shall be mainly concerned in this chapter with the judicial side, aka *research ethics*.

1.1. Legislative knowledge policy

Legislative social epistemology is most vividly understood in terms of "turning points" in history. The question it answers is "where do we go from here?" A significant version of this question contributed to the pre-history of Kuhn's *The Structure of Scientific Revolutions* (Fuller 2000b: ch. 2). In the decade prior to World War I, the German scientific community debated the future of physics at a time when the Newtonian paradigm was at the height of its success, but it still could not provide a satisfactory account of the nature of light. Yet no consensus had been reached over what by 1920 had consolidated into the new Einsteinian orthodoxy. From today's standpoint, the one that Kuhn himself adopts, the decision that had to be made was clear: the dominant paradigm must eventually yield to a successor capable of resolving its persistent problems. In the original debate, this position was epitomized by Max Planck, the editor of the physics journal that first published Einstein. Planck and Kuhn presumed that the autonomous pursuit of science is always an end in itself. However, the alternative viewpoint was exemplified in Planck's day by another physicist, Ernst Mach, whose "instrumentalist" approach to science has been more influential outside than inside his home discipline. Rather than indulge in greater specialization, Mach believed that scientists in a mature paradigm should forsake intractable problems of

limited technical interest and instead apply their knowledge to ame-
liorate the human condition.

In short, Planck wanted more science producers, and Mach
more science consumers. If one were to contrast Planck's and Mach's
"science utility functions," as philosophers from Charles Sanders
Peirce to Nicholas Rescher would advise, Mach's approach would
stress the diminishing returns on additional investment of effort in
the dominant paradigm, whereas Planck's would counter with the
increasing value of each such return, making the additional effort
worthwhile (Rescher 1978). What Mach derogated as esoteric tech-
nicalities, Planck valorized as the deepest truths. Their disagreement
paralleled one of national economic policy that began around the
same time and came to dominate political debate in the 20th century:
how does the state maximize social welfare – by distributing through
taxation the surplus of wealth already produced (cf. Mach) or
concentrating investment in innovative capital development schemes
(cf. Planck)?

Corresponding to "distribution" in the science realm is the rou-
tinization of technical knowledge as easily available technology that
enables people to pursue their ends without first having to acquire a
costly expertise. The result is to disaggregate belief from utility: the
same tools are usable by people whose beliefs may radically differ.
For the Austrian school of economics that pioneered the economics
of knowledge in the middle third of the 20th century, this was the
preferred way to conceptualize knowledge as "public good" (e.g.
Machlup 1962). For Mach, a spiritual godfather, it was the ultimate
realization of the Enlightenment dream of "science in a free society,"
to quote the book title of his most celebrated recent admirer
(Feyerabend 1979). In contrast, Planck would bind belief and utility
more tightly together, turning science into a world-view that would
colonize and reorganize everyday life, leading to a greater unity of
purpose, whereby the whole of humanity might achieve more than the
sum of its particular individuals. Scientization – what Max Weber
called "rationalization" – would thus bring to fruition the promise of
imperialism, a British political invention that the Germans sorely
wished to match. From this standpoint, Lenin's fascination with
imperialism as providing the infrastructure for revolutionary socialism, like Comte's fascination with Roman Catholicism as the institutional vehicle for global positivism, pointed Marxism toward a
Planck-like future, which was expressed most clearly in his anath-
ematization of Russian followers of Mach in *Materialism and Empirio-
Criticism* (1908).

It would have been interesting to see Mach's and Planck's reaction to the emergence of a natural language (English) as the universal medium of scientific communication. Until the end of World War II, when the US became the undisputed leading world power, it was widely believed – certainly by Mach's spiritual heirs, the logical positivists – that any multipurpose scientific medium would have to be either a made-to-order scientific language, that is, a kind of interdisciplinary Esperanto, or an advanced form of logic that transcends the syntactic and semantic differences of natural languages. Selecting a natural language for such a task would run the risk of turning the language into a hegemonic world-view, something Planck invited but Mach abhorred. Keep in mind that the likely candidate for this hegemonic role in their day was not English, but German, whose defenders and opponents both stressed the ideological freight the language carried in its syntax and semantics.

One aspect of the Mach–Planck dispute bears specifically on proprietary concerns about scientific knowledge. Mach treated intellectual work as merely a high-skilled form of industrial labor, whose relief through technological innovation should be expected and welcomed. Like the champions of entrepreneurship and management of the period, Henry Ford and Frederick Winslow Taylor, Mach regarded the protected status of any form of labor as a vestige of the guild mentality that impeded the free exchange of goods, services, and – as far as Mach was concerned – ideas. The Platonic precedent of treating mental activity as the monopolistic preserve of an elite is, to recall George Bernard Shaw's contemporaneous swipe at experts, a "conspiracy against the public interest." On this basis, Mach singled out the Athenian philosophical elite – Aristotle in particular – for having retarded scientific progress. Not only did their alienation from manual labor engender false physical ideas, but their literacy, the most durable means of knowledge transmission, also enabled those false ideas to be spread more widely than those of their illiterate slaves, who held intuitively more correct ideas about matter and motion (Fuller 2000b: 115).

Had Mach lived longer into the 20th century, he would have probably celebrated the rise of electronic calculators for rendering mathematical knowledge more widely available and reducing the proprietary hold – not to mention mystique – that professional mathematicians exert over their skills. However, Mach would not have anticipated the conversion of calculating devices, and then computer programs, into intellectual property as an ever expanding domain of mathematical objects conspired with an ever expanding patent system. Despite his sensitivity to the need to democratize knowledge, Mach saw only the "destructive," not the "creative," side of an innovation's creative destruction of the market. He failed to see how innovation often reconfigures, rather than eliminating outright, the difference between the "haves" and the "have-nots."

1.2. Judicial knowledge policy: aka research ethics

The judicial side of social epistemology is the focus of the rest of this chapter. Let us begin with an observation: science, and the behavior of scientists, has never been perfect, but now the imperfections seem to matter more. Consider the South Korean scientist Hwang Woo-Suk, who was selected as one of *Time* magazine's "People Who Mattered" in 2004, for having allegedly extracted stem cells from cloned human embryos. Nevertheless, despite having published in *Nature* and *Science*, the world's leading scientific periodicals, Hwang's research was found to be based on unethical treatment of subjects and, more importantly, the misrepresentation of data (especially multiple reportings of the same data). By Christmas 2005, Hwang had been forced to resign from Seoul National University, and six of his co-workers were suspended or had their pay cut. Hwang and his colleagues are hardly alone. In response to the recurrence of well-publicized and highly damaging scandals across the physical and biological sciences in recent years, many universities and some entire national research funding agencies now convene "institutional review boards" to deal with breaches of what has come to be known as "research ethics."

It may be flippant to regard the recent emergence of "research ethics" as an academic craze. But perhaps "moral panic" is a more apt description. To be sure, from artists to scientists, all intellectual workers are preoccupied with the giving and taking of credit. The hiring, promoting, and rewarding of academic staff is increasingly based on "citation counts," which consist in the number of times someone receives credit in peer-approved publications. Even if someone's work is criticized, it must be credited properly. Often credit is given simply because one has already produced creditable work (this is common in grant funding decisions and can even apply in publication decisions, in terms of giving the benefit of the doubt). But increasingly the fixation on credit reflects the work's potential monetary value. In the ever-expanding field of intellectual property law, the former is covered under copyrights, and the latter under patents. However one defines the creditworthiness of intellectual work, one thing is clear: reported cases of fraud are on the rise.

So perhaps there is, after all, a need for national research ethics boards or even an international court of scientific justice. If so, what form should these institutions take? These questions belong to the domain of *epistemic justice*, which attempts to solve the problem of how knowledge may be produced so as to result in the distribution, rather than the concentration, of power. In the slogan "knowledge is power" (or "*savoir est pouvoir*" or "*Wissens ist Kraft*"), power involves *both* the expansion and contraction of possibilities for action. Knowledge is supposed to expand the knower's possibilities for action by contracting the possible actions of others. These "others" may range from fellow knowers to non-knowing natural and artificial entities. This broad understanding of the slogan encompasses the interests of all who have embraced it, including Plato, Bacon, Comte, and Foucault. But differences arise over the normative spin given to the slogan: should the stress be placed on the *opening* or the *closing* of possibilities for action? If the former, then the range of knowers is likely to be restricted; if the latter, then the range is likely to be extended. After all, my knowledge provides an advantage over you only if you do not already possess it; hence, knowledge is a "positional good" (Hirsch 1976).

This idea of positional good also helps to explain the rather schizoid attitudes toward the production and distribution of knowledge that are epitomized in the constitution of universities. In short, we do research to expand our own capacity to act, but we teach in order to free our students from the actions that have been and could be taken by others. I have characterized the university's management of this schizophrenia as "the creative destruction of social capital" (Fuller 2003b). By virtue of their dual role as producers and distributors of knowledge, universities are engaged in an endless cycle of creating and destroying "social capital," that is, the comparative advantage that a group or network enjoys by virtue of its collective capacity to act on a form of knowledge (Stehr 1994). Thus, as researchers, academics create social capital because intellectual innovation necessarily begins life as an elite product available only to those on "the cutting edge." However, as teachers, academics destroy social capital by making the innovation publicly available, thereby diminishing whatever advantage was originally afforded to those on the cutting edge. In this respect, intellectual property is anathema to the very idea of the university (Fuller 2002a: ch. 1). Recalling Joseph Schumpeter's (1950) definition of the entrepreneur as the "creative destroyer" of capitalist markets, the university may be regarded as a "meta-entrepreneurial" institution that functions as the crucible for larger societal change.

If universities are the ideal dispensers of epistemic justice, then what would be the exact charge of a national research ethics board in the case of, say, Professor Hwang? In other words, what is the nature of the problem associated with his research – that he claimed personal credit for work not done, or that he allowed that work to be disseminated widely to other medical researchers and practitioners? The former bears more directly on the potential financial benefits from cloned human stem cells, while the latter bears on the potential consequences for recipients of treatments based on such stem cells. The ambiguity here immediately points to a problem in defining the extent of researcher responsibility. One nation has already taken significant steps in the right direction to address the systemic issues suggested here: Norway, whose code of research ethics – available at <http://www.etikkom.no> – was drafted by a team of academic lawyers.

The key feature of this code is its comprehensive coverage of the various conditions under which the researcher's autonomy is challenged: from government, business, administrators, colleagues, subjects, stakeholders, and so on. In each case, reciprocal rights and duties among the parties are identified for legal protection. The overall image produced by the Norwegian code, historically fostered by the German legal tradition, is that academic freedom amounts to security of the researcher's sovereignty over a domain of activity (i.e. "freedom" as the literal translation of *Freiheit*). The stress, then, is on the setting of boundaries between, say, where the stakeholder's domain ends and the researcher's begins. To be sure, these boundaries may change over time, but at any given point they must be explicitly defined so that the autonomy of what lies on both sides of the divide is maintained.

So far we have defined the "essential tension" of research ethics in terms of maintaining a delicate balance between crediting and publicizing research. Violations of this balance can be generically classed as *fraud*. However, as we shall see in the rest of this chapter, the negative status accorded to acts of fraud is highly context dependent. We shall proceed, first, by exploring what might be called the "metaphysical" foundations of research fraud: how must the world (including ourselves) be constituted for the idea of research fraud to make sense? Then we turn to the Lomborg Affair, a recent high-profile case of alleged fraud in the broad sense of misrepresenting the scientific work of oneself and others. This will give us an opportunity to examine competing regimes of epistemic justice. The chapter concludes with some reflections on how historical consciousness can wreak havoc on our intuitions about research ethics.

2. Research Fraud: Its Theoretical and Practical Sides

2.1. Defining fraud: crime or learning experience?

Minus the moral freight, the commission of "fraud" on another or oneself (i.e. deception and self-deception), used to carry serious epistemic weight in the social sciences. From the anthropologist Lucien Lévy-Bruhl to the developmental psychologist Jean Piaget, the early 20th century empirical study of irrationality largely centered on false notions of causation possessed by, say, primitives or children. There were two general strategies for demonstrating irrational beliefs involving causation: one induced by the investigator and the other inferred by the investigated. The former consists of *manipulating* the environment of the investigated so that they think a natural cause was behind a phenomenon that the investigator has artificially constructed. If the manipulator regularly succeeds, then the manipulated demonstrates an immature state of cognitive development. The latter strategy consists of *superstitious* causal beliefs that the investigated spontaneously form from observing the close correlation between an event or activity and some target phenomenon. This is something an ethnographer might observe over time in native behavior or detect in the narratives of their tribe's exploits. If manipulation is about deception, superstition is about self-deception.

Forgers and *plagiarists* manufacture situations in the research world comparable to that of manipulation and superstition, respectively. Like manipulators, forgers create the impression that something or someone else has produced an artifact that is in fact of their own design. In contrast, plagiarists capitalize on the superstitious tendency to infer causation from correlation – in this case, that the juxtaposition of a name and a text implies authorship of a piece of research.

When officially incorporated as part of the research process into human subjects, susceptibility to manipulation and superstition has historically distinguished the rational (researcher) from the irrational (researched). Clearly, the distinction is meant to set the terms of epistemic authority such that the former benefits at the expense of the latter. However, when unofficially applied to the researchers themselves as forgery and plagiarism, displays of manipulation and superstition are eligible for moral censure. Moreover, it is the perpetrator, not the target of perpetration, who is censured for having committed an act of fraud. (After all, it is easy to imagine a possible world in which those fooled by forgery and plagiarism would be blameworthy and forced to live with the consequences of their misapprehension.) Why the distinction – indeed, the inversion of epistemic and moral judgment – between the normal and the reflexive contexts of inquiry? Perhaps without such a distinction, however arbitrarily enforced, it would not be possible to circumscribe the sphere of science from art in its purest sense (cf. Grafton 1990). In any case, the concept of fraud is a good place to begin.

Fraud comes in as many forms as credit. Sometimes fraud involves taking credit for someone else's work. But often, especially in more competitive fields of science, fraud takes the exact opposite form: someone or something else is given credit for something the culprit has fabricated. The former fraud is associated with plagiarism, the latter with forgery. The one constitutes a sin of knowledge distribution, the other of knowledge production. In neither case is the competence of the fraudster in doubt. The successful plagiarist and forger must be the intellectual equal of those whose work they reproduce or anticipate, and perhaps the intellectual superior of the audiences on whom their success is predicated. The interesting question, then, is why these forms of fraud are treated with increasing moral outrage.

But before we address this question, first note that the crediting of research is only one of several possible functions performed by the *citation* of research. By "citation" I mean the simple act of providing readers with information about where they can learn more about what the author says and the context in terms of which she would have her claims understood and evaluated. This may or may not involve referring to works that were actually material in the text's composition. Indeed, an author whose intellectual resources far exceed those of her readers may even make a point of citing readily accessible works whose relevance to the author's conceptualizations may be negligible. Much textbook writing still has this character, and it was probably the spirit in which references were usually made before citation came to be read as the bestowal of credit (Grafton 1997).

An interesting test of the normativity of citation for both credit and information is the explanation that the maverick early 20th century economist Thorstein Veblen reportedly gave for why he rarely cited anyone: it would insult the expert and intimidate the lay person. In other words, the reader would either already know the source for his claims or she would be in no position to evaluate it. For Veblen, whose legendary lack of collegiality was the dark side of his attentiveness to the "intelligent public" spared his bibliography, this was a formula to write in a self-sufficient manner that can be understood and judged without relying unduly on the author's "connections."

From the logic of "surplus value," plagiarism is easier to understand than forgery. Whereas the successful forger must possess much of the same technical expertise as that which she forges, plagiarism typically requires skills that are less costly to acquire and execute. The latter rides the coat-tails of a garment the former could have designed. Thus, it may be said that when a student plagiarizes a term paper for a course, or an author plagiarizes the writings of a fellow author, the plagiarist attracts opprobrium for her inappropriate sense of efficiency. Nevertheless, the plagiarist possesses a valuable skill, one comparable to connoisseurship. The plagiarist makes a calculated guess about what will have the desired effect on the target audience. She may not know how to write a good article on Nietzsche, but she knows one when she sees it – a skill she then uses to get the article published under her own name.

The pedagogical value of her achievement should not be underestimated. In contrast, there is the person who can write articles on Nietzsche good enough to pass peer review for publication, yet lacks any clear sense of the difference between the good and the bad stuff written on Nietzsche – often because she can only relate what others write to her own work. Which of these two people is better trained? I would say clearly the former, the plagiarist. She has understood the spirit of Nietzsche scholarship, for which she then unfortunately takes undue credit, whereas the latter has merely submitted her mental life to the discipline of writing passable Nietzsche articles. Which will it be then: the fair-minded plagiarist or the scrupulously honest scholar who rejects anything that does not bear her own handiwork?

Shorn of gratuitous moralism, the plagiarist's case can be dealt with summarily in purely epistemic terms. On the one hand, if the plagiarist succeeds in passing off someone else's work as her own, then she has effectively taught her evaluators something *they* did not already know. To quibble with this conclusion (and quibble they do) is to promote the absurd idea that candidates for degrees or publication should be held to a *higher* standard than those who would judge them. On the other hand, if the plagiarist is caught, her competence can be tested in an oral examination, whereby, say, the plagiarist is required to discriminate between pairs of texts on, say, Nietzsche scholarship in terms of their desirability as targets for plagiarism. This would be like distinguishing the lucky from the truly knowledgeable on multiple-choice exams by making the options increasingly hard to distinguish: the probability that an incompetent could guess right would be very low. With Alan Turing's blessing, the plagiarist would count as competent if her lucky guesses manage to coincide with those who are known to have undergone the requisite mental discipline.

Perhaps the reader will find my approach to fraud too "postmodern," redolent of Jean Baudrillard's (1983) philosophy of hyperrealism, whereby the simulation sets the standard for the real. Thus, it would seem, I would have the plagiarist dictate what passes as genuine competence by measuring how well she can pass herself off as something she is not. There is considerable truth to this observation, but it sounds much more radical than it really is, since it is already the normal practice of peer review journals, which typically do not visit sites of knowledge production but make surface validity judgments – engaging in what Steven Shapin (1994) has called "virtual witnessing" – of articles submitted for publication (Fuller 2006a: 54–8). What follows from this practice? Not that there is relatively little fraud but that fraud matters relatively little.

If the reader remains disoriented, then she begins to appreciate the quandary in which 18th-century thinkers found themselves once they took seriously the idea, associated with Bernard Mandeville and later Adam Smith, that "private vices make for public virtue" (Hirschmann 1977). Only wishful thinking could lead one to conclude that morally better people make sounder business partners and, similarly, cognitively superior scientists. What matters in both cases is that the appropriate institutions are in place to ensure that people's natural moral and epistemic liabilities are held in check so that their common product is always greater than the sum of their individual interests (cf. Rescher 1978). This is the true genius behind constitutional design (Fuller 2000a). It marks a crucial moment in "secularization," in that people of very diverse motives and interests agree to abide by the same set of externally administered standards. In exchange for this level of procedural conformity, individuals can rest assured that their souls will not be further inspected.

So, is there anything more to the "objectionable morals" of the plagiarist than her evident contempt for hard work as an end in itself? The plagiarist would find a defender in Karl Marx's son-in-law, Paul Lafargue, who proposed a "right to be lazy" (Pachter 1984: ch. 1). Lafargue worried that Marxists ran the risk of fetishizing work, losing sight of new technology's long-term capacity to relieve drudgery and expand the sphere of leisure out of fear of technology's short-term tendency to dehumanize the labor process and deprive workers of income. A century later, the Sartrean André Gorz (1989) a little too hopefully called on Marxists to prepare for a high-tech version of Lafargue's post-industrial future. More generally, the *Lafarguiste*

strand in the history of socialism has fallen foul of capitalism's seemingly endless capacity for rejuvenation. Nevertheless, it has forced
those valuing sheer production over the optimal use of resources (i.e.
true "productivity") to justify their preference.

This burden weighs most heavily today on postmodern purveyors
of both pragmatism and identity politics. For them, in the wake of
globalized exchange patterns, the loss of not only such explicitly economic entities as native industries and guild practices, but also deeper
biosocial ones like local languages and genetic distinctness, is tantamount to annihilation, since what people are is, quite literally, what
they produce (for themselves) and reproduce (of themselves). When
people start doing otherwise, they become someone or perhaps even
something else. In other words, all of what Lafargue and his followers treat as "drudgery" is, in fact, essential to who we are. Behind this
conclusion is an assumption that surfaces in contemporary debates
over intellectual property: namely, that effort which results in a thing
of value – including the conduct of a meaningful of life – is a risky
proposition with a low prior probability of success. This intuition goes
back to the sacred nature of creation, and the correspondingly profane
nature of imitation, which recurs as a theme in the more puritanical
strains of Judaism, Christianity, and Islam. It is associated with "iconoclasm," the principled destruction of representations of the divine
lest they be worshipped in a manner reserved for the original deity.
The proliferation of easily available images degrades the deity by
removing the beholder's need to affirm God's originality.

Such iconoclasm is apparent in Marshall McLuhan (1964), a practicing Catholic who in earlier centuries would have been judged
heretical, as he provided an intellectual basis for challenging the
Church's generally "iconodule" (icon-worshipping) disposition. He
classified television as a "cool medium," implying that viewers must
make a special effort to overcome the cumulative effect of the
medium's *prima facie* clarity of representation. The Frankfurt School
critic, Walter Benjamin, partly anticipated this complaint when he
wrote of a loss of "aura" that results once an identity becomes
mechanically reproducible to such an extent that its originator no
longer controls access to it (Benjamin 1968). This aura, historically
associated with both divine originality and the soul's privacy, could be
always recovered in religious cultures with a cultivated sense of the
supernatural, since the soul would ultimately escape all attempts to
control its material expressions (indeed, Epictetus the former Roman
slave founded Stoicism on precisely this basis). However, the super-
naturalist option is clearly unavailable to secularists, who must then

seek simulations. Unique memory traces, historical trajectories, and artifacts thus become markers of personal identity. In this respect, art functions for Benjamin as labor did for Marx: the simulator of the sacred in a secular world.

From this perspective, it makes sense to heighten the sanctions surrounding forgery and plagiarism, as signified by the excessive care lavished on artworks – but also the increasingly sophisticated authentication techniques applied to less exalted products of the human imagination, like student essays. The diametrically opposed viewpoint, one that gladly profanes the soul by depicting it as a forged and plagiarized pastiche of previous lives, is represented by Daniel Dennett's (2003) pan-Darwinist, meme-driven theory of personal identity.

In the next section, I turn to one of the most intellectually interesting and politically relevant cases of alleged research fraud in recent times, the Lomborg Affair. However, it is first worth noting just how circumscribed our discussion of fraud has been up to this point. We have so far assumed that, whatever the ambiguities surrounding the motives and actions of those accused of fraud, their alleged fraud pertains to works that can be determinately classified as either fact or fiction. But what if there is ambiguity even at this basic level? How does the problem of determining whether a text is fact or fiction bear on whether it is original or plagiarized? This question was at issue in the highest profile trial held in London's High Court in early 2006. It pitted two parties, both having done well by their common publisher, Random House. The plaintiffs were two of the three authors of *The Holy Blood and the Holy Grail*, a 1982 best-seller that offers a speculative history of the personal life of Jesus Christ, in which the Resurrection is interpreted to mean that Jesus did not die on the Cross but lived to wed the prostitute Mary Magdalene. Their descendants settled in France, married into a royal dynasty and are now supposedly set on repopulating the thrones of Europe. The defendant, Dan Brown, was the author of an even bigger best-seller, *The Da Vinci Code*, a novel first published in 2003, which has sold 40 million copies in over 40 languages – and is the subject of a major motion picture starring Tom Hanks. Brown was alleged to have stolen the theory behind the plot of *The Da Vinci Code* from the authors of *The Holy Blood*.

Both sides of this case offered provocative arguments that provide the basis for a major reconceptualization of the relationship between thinking and owning. However, the books on trial officially belong to different literary genres, though the conflict between them reminds us that the two genres they represent were born joined at the hip. *Don Quixote* by Miguel de Cervantes, normally regarded as the first novel, is written as an account of an Arabic manuscript discovered in a bazaar. However, it was clear from its original publication in 1605 that *Don Quixote* was fiction presented as history. The speculations advanced in *The Holy Blood* may be fiction as well, but its authors claimed to be unable to falsify them. This rather "postmodern" attitude contrasts with *The Da Vinci Code*, which is much more like *Don Quixote* in that readers are never in doubt that they are dealing with fiction. This, in turn, justifies the liberties that Dan Brown takes with the plot, since in the end he is trying to tell a compelling story rather than to report what some people believe might be true. Nevertheless, it is also clear that Brown had read *The Holy Blood* when composing *The Da Vinci Code*. Indeed, the name of the novel's villain is an anagram of the names of the earlier book's authors.

The baseline of copyright law (at least in the UK, where copyright originated) is that copyright does not extend to ideas as such but only to their specific embodiment. So, the burden on the plaintiffs was to show that Brown could not have written *The Da Vinci Code* without having read *The Holy Blood*. It is easy to see why, in today's world, this provision would have the ironic consequence of generating more lawsuits, yet litigants would find it increasingly difficult to prove their case. On the one hand, it has become customary, if not required, that authors cite all the sources for their work, and these sources have become more readily available for inspection. On the other hand, just this easy availability of sources makes it harder to establish the uniqueness of the author's need for a specific source, since the author could have easily relied on an alternative source. Thus, while Brown draws on a historical speculation presented in *The Holy Blood*, that book was not the only source at Brown's disposal that proposed the idea – at least given the state of Brown's primary reference tool, the Internet, at the time he wrote *The Da Vinci Code*, two decades after the publication of *The Holy Blood*. The plaintiffs had to show that there is something about the arrangement of the alleged events recounted in *The Holy Blood* – its "narrative logic" or "theory" – that Brown could not have obtained elsewhere. But this too is hard to prove because, as a novel, *The Da Vinci Code* also deviates significantly from the plot of *The Holy Blood*.

That Brown won the case – and the attendant £1.3 million court costs – in a short judge-based trial is perhaps unsurprising. Despite Brown's much greater financial success, his opponents were portrayed unsympathetically in the media (and even by the judge) as greedy rent-seekers whose victory would have had a chilling effect on free expression. Nevertheless, the lawsuit raises two interesting issues. First, the plaintiffs would not have had a case at all if their historical speculation had been verified, since accepted truth is in the public domain. It was to the advantage of the authors of *The Holy Blood*, if only in terms of enabling litigation, that their hypothesis was not yet widely regarded as true. Should some decisive evidence have been uncovered during the trial that confirmed their speculation, the case would have immediately collapsed: their claims would have passed from the realm of fiction, which is subject to intellectual property claims, to that of fact, which is not. Of course, at the same time that the authors of *The Holy Blood* might lose the right to accuse Brown of plagiarizing their narrative logic, they may become eligible for a major prize rewarding historical investigation.

2.2. The Lomborg Affair: fraudster or scapegoat?

After this philosophical survey of the different senses and contexts in which a charge of "research fraud" might be raised, let us now focus on a single case. The administration of epistemic justice is instructively raised in the context of a negative example I witnessed as a visiting professor in the Copenhagen Business School. In January 2003, the world learned that Denmark's national research council convenes a body whose name is rendered ominously in English as "Committee of Scientific Dishonesty" (CSD). The CSD normally deals with forms of misconduct – negligence and fraud – that beset the more competitive reaches of the biomedical sciences. However, the CSD had now set its sights on Bjørn Lomborg, a young associate professor of political science at the University of Aarhus, whose transatlantic bestseller *The Sceptical Environmentalist* purported to show that ecologists routinely overstate the world's environmental problems to fit their political agenda (Lomborg 2001). If the CSD thought it would settle the matter once and for all, it was mistaken. Shortly after the committee's judgment, the Danish government commissioned an inquiry into the CSD's own future.

The Sceptical Environmentalist has been subject to intense public scrutiny since its British publication in late summer 2001. Some hailed Lomborg's honest reappraisal of ecology, a field arguably shrouded in political correctness, while others accused him of malicious incompetence in the use of data. None of this has hurt the book's sales, despite its 500 pages and 3,000 footnotes: five years after publication, *The Sceptical Environmentalist* remains in the top

1,000 best-sellers on offer by Amazon, the on-line book vendor, on both sides of the Atlantic. Support and opposition for *The Sceptical Environmentalist* have divided along predictable lines, with, say, *The Economist* championing Lomborg and *Scientific American* condemning him. Somewhat more surprising is that the main parties to the dispute have been either just as removed from the relevant front-line natural science or as politically motivated as Lomborg himself. A naïve observer could be forgiven for concluding that Lomborg had violated an implicit norm of public ecological discourse: "If you're going to be political, make sure you're a certified environmental scientist. Otherwise, toe the line and support the ecologists!"

As it happens, the author of *The Sceptical Environmentalist* claimed he spent only 18 months working with students locating errors and distortions in the statistics used by ecologists to fuel their gloomy forecasts. Lomborg had not previously made any notable first-order contributions to environmental science or policy, though his formal training in rational choice theory and the political economy of welfare informed the book's optimistic case for the future of humanity. However, Lomborg's fiercest public opponent, Harvard's E. O. Wilson, was himself no more qualified than Lomborg to pass judgment on the empirical state of the global environment, though he certainly has voiced informed lay opinions on the matter (e.g. Wilson 2002). The professional reputation of Wilson, best known as the founder of sociobiology, rests on his studies of the social behavior of ants. Methodologically speaking, Wilson is a traditional natural historian, not especially expert on the interpretation of statistics, on which charges of Lomborg's incompetence are largely based. Wilson's significance in the controversy surrounding *The Sceptical Environmentalist* – he spearheaded the attack on Lomborg in *Scientific American* that helped to justify the CSD's inquiry – puts paid to the idea that politics is never the trump card in science. Wilson had spent much of the past 30 years vilified for the political and scientific malfeasances allegedly committed in the name of sociobiology, but now, it would seem, the field has gained renewed scientific respectability as "evolutionary psychology," and Wilson himself has been resurrected as a heroic ecologist, responsible for coinages like "biophilia" and "biodiversity" (Wilson 1984, 1992).

Aside from the pressure exerted by international notables like Wilson, Lomborg was stigmatized by the Greens and the recently ousted Reds of his own country for having benefited from a newly elected neo-liberal government, which made him director of an environmental assessment institute with an eye to rolling back pollution

restrictions on industrial growth. To be sure, as Denmark's highest
profile representative at June 2002 Johannesburg "Earth Summit,"
Lomborg argued for the loosening of such restrictions. At the same
time, however, he called for the profits from increased industrial
output to be channeled into development aid to third world nations.
This old-style leftist sentiment surprised Lomborg's neo-liberal
paymasters. Indeed, it showed that the controversy surrounding
The Sceptical Environmentalist turned on something deeper than
Lomborg's scientific competence. There loomed the larger question
of who is authorized to define the field of environmental science.

At this point, let us examine *The Sceptical Environmentalist*'s table of
contents:

The table makes clear that Lomborg would have the field of environmental science subsumed under welfare economics, such that the natural environment is valuable only insofar as it contributes to the greatest good for the greatest number of humans, not as an end in itself. This point came out even more strongly in Lomborg's highly publicized gathering of world-class economists in 2004 known as the "Copenhagen Consensus" (Lomborg 2004). In contrast, Wilson and most professional environmental scientists see themselves as concerned with the future of all forms of life. Thus, they would judge humanity by our ability to conserve the natural environment. Whereas Lomborg holds that biodiversity may need to suffer for human starvation to be eliminated, his opponents' unqualified support for biodiversity extends to recommending curbs on human population growth. In their own way, each side is calling for "sustainable development," but the difference in how they order priorities leads to substantially different development policy proposals. In many respects, these differences replay the original debates over the ecology that erupted in the late 1960s. However, back then, the difference of opinion was not so sharply divided along disciplinary lines: specifically, there were biologists on *both* sides of the issue, with Lomborg's position not so very different from that of Barry Commoner, who also saw the imposition of population checks on the poor as a politically cheap way of tackling the problem of the inequitable distribution of wealth (Fuller 2006b: ch. 13; cf. Schwartzman 1995).

A close look at Lomborg's re-analysis of environmental data reveals someone cautious about recommending draconian policy solutions on purely biological grounds. He fully realizes that such solutions are designed to curtail not only the profligate consumption patterns of the rich (which attracts media coverage) but also the profligate reproductive patterns of the poor. For example, in chapter 6, when considering whether the gap between the world's rich and poor is truly increasing, Lomborg observes that the United Nations Development Program nurtures such fears by drawing attention to the fact that poor families out-reproduce rich ones. Yet, Lomborg notes, the purchasing power of poor families is increasing (as measured by the percentage of income needed to supply essential goods). To be sure, this point does not justify complacency, but it does suggest that century-old eugenics policies need not be part of a comprehensive package to reduce global poverty. If the right metrics are used – ones appropriate to the phenomena that need to be tracked – then there is still reason to back now unfashionable welfare policies that would encourage economies to grow themselves out of poverty.

Later on, in Part V, Lomborg tackles the presuppositions of Wilson and his allies more directly. Here he immediately grants that biodiversity must be part of a comprehensive welfare policy, in that no nation should ever be dependent on a single staple crop for its economic support. However, Lomborg is suspicious of defenses of biodiversity that go beyond this minimalist thesis because they tend to be based on a peculiar ideological reading of neo-Darwinism. For, if Darwin is correct that there is always a struggle for survival because species' reproductive patterns tend to outstrip available resources, then extinction is simply a natural biological fact. Extinctions that are specifically traceable to human activities – insofar as these can be clearly established – should be considered in that light as normal. Humans do what they can to maximize their survival, just as other species do, often to the detriment of their biological co-habitants (the spread of micro-organisms that curtail human populations in the guise of "epidemics" and other "diseases" comes to mind). Whatever normative claim is made for maximizing the diversity of species on the planet goes beyond Darwinian strictures. Moreover, the norm remains difficult to promote because of lingering uncertainty over the actual number of species that have inhabited the planet. What, then, underwrites the normative force of biodiversity? It may be an anthropomorphic transfer of egalitarianism from an intra-species value among humans to an inter-species value as such. Perhaps this represents a mutant version of the Christian theological ideal of humanity's stewardship of Earth. At the same time, biodiversity may also be rooted in misanthropy, since most biodiversity policies call for the curtailment – eugenic or otherwise – of humanity's presence on the planet.

Lomborg pursues this last prospect at the end of the book, when considering global warming, the source of his most controversial interventions. The issue here is not the sheer phenomenon of global warming, which virtually everyone grants – but its significance, especially in terms of the kind of policy problems it poses. The very name "global warming" hides potential confusion, as it allows for solutions that may be, in some abstract sense, globally optimal (e.g. overall reduction of the world's carbon emissions), yet suboptimal with respect to parts of the globe in relatively weak bargaining positions. In this respect, the concept of global warming is in the grip of the idea of Earth as a superorganism, which James Lovelock has christened "Gaia." One thinks here of the recent popularity in markets for "carbon shares" that can be traded, allowing debt-ridden poor but clean countries to have their environments polluted by rich countries that have exceeded their normatively desirable carbon emissions levels. Still more pervasive is "corporate environmentalism," whereby industries adopt eco-friendly practices that involve the exploitation of human labor, if not its outright unemployment (Hoffman 1997). Such global strategies neglect the basic fact of political economy that the ability of a society to adapt to environmental changes depends on its levels of wealth and innovation. That global warming harms poor societies disproportionally simply reflects the general inability of the poor to control their fate, regardless of its source. Instead of aspiring to Gaia's speculative standard of superorganic equilibrium, would it not be better, then, to foster humanity's flexibility through economic growth policies?

The intellectual location of environmental science turned out to be decisive for activating the CSD. *The Sceptical Environmentalist* appeared to be a book on environmental science that was never formally vetted by a "proper" environmental scientist. It was published by the social science division of Cambridge University Press. This suggested the kind of false advertising that might be associated with "scientific dishonesty." It raised the specter of the so-called Sokal Hoax of 1996, whereby a disgruntled physicist managed to publish a scientifically nonsensical but politically correct article in a leading cultural studies journal. The perpetrator of the hoax, Alan Sokal, claimed to have shown that without peer review, some trendy humanists will publish anything that confirms their prejudices about science (Sokal and Bricmont 1998; for critical philosophical assessments, see Babich 1997; Fuller 2006a: ch. 4).

However, in the Lomborg case, the question of who was hoaxing whom remained open. The Danish title of his book – literally translated as *The Real State of the World* – explicitly parodied the Worldwatch Institute's earnest and resolutely gloomy annual report, *The State of theWorld*. Moreover, Lomborg never claimed to offer original evidence for the issues surrounding the state of the environment. Rather, he reinterpreted the statistical evidence already in the public domain. Lomborg claimed to have shown first the looseness of fit between the indicators and what they are supposed to indicate, and second the ease with which those indicators can be made to point to a much more benign prognosis for the human condition.

The CSD eventually found Lomborg guilty of having produced a work that had the effect of misleading readers without necessarily intending to mislead. No one was satisfied by this decision, which was quickly appealed. The CSD had clearly failed to establish the grounds for its own authority. Both Lomborg's supporters and opponents felt vindicated in their original positions, which only served to polarize, not reconcile, the parties. In particular, Lomborg's defenders questioned the CSD's fairness in singling him out for rebuke when the most prominent environmental scientist opposing him, Stephen Schneider, confessed that "we need to get some broad-based support to capture the public's imagination. That, of course, entails getting loads of media coverage. So we have to offer up scary scenarios, make simplified, dramatic statements, and make little mention of any doubts we might have . . . Each of us has to decide what the right balance is between being effective and being honest" (*Economist* 2002). Fair enough. But then why did the CSD agree to judge a case in which political motives and scientific practice were so intimately entangled?

Not surprisingly, the Danish research council has radically reorganized the CSD, reflecting a widespread retrospective feeling that the CSD had overstepped its remit by trying to censor legitimate public disagreement about the place of science in society. Perhaps the CSD should have simply refused to judge the Lomborg case. I agree. At the time of the Lomborg controversy, the CSD was an unholy mixture of two models of judicial review that are capable of authorizing a research ethics board. On the one hand, the CSD has elements of a proactive *inquisitorial* system that promotes an independent standard of scientific propriety in terms of which many scientists may be found generally wanting. On the other hand, the CSD resembles a more reactive *accusatorial* system that presumes scientists innocent of impropriety unless a formal charge is brought against them. (I originally applied this distinction from the law of evidence to the philosophy of science in Fuller 1985: ch. 1, n. 28.) I summarize the characteristics of the two systems as alternative visions of epistemic justice in Boxes 5.1 and 5.2 below.

The natural context for an inquisitorial system is a field whose scientific integrity is regularly under threat because its research is entangled with political or financial interests. Often these entanglements are unavoidable, especially in the case of biomedical research. Here

---

Box 5.1 The inquisitorial approach to epistemic justice

- Motivated by the idea that the interests of "science" are independent of scientists' interests, and that scientific error will not naturally self-correct.
  - Scientists presumed guilty until proven innocent.
  - Apt for research whose integrity is easily (systematically?) compromised by political and economic considerations.
  - Allows the agency to raise enquiries of its own when no one has filed a complaint.
- Requires a clear sense of jurisdiction to establish a pretext for inquiry.
  - Was the agency involved in enabling the action (e.g. funding)?
  - Can the agency enforce sanctions based on its ruling (e.g. excommunication)?
- Respected for value-added mode of inquiry that promises closure without simply reinforcing the status quo.
  - Independent examinations of affected parties and witnesses.
  - Explicit demarcation and weighting of existing evidence.
  - Restitution is a collective learning experience for science.

Box 5.2 The accusatorial approach to epistemic justice

- Motivated by the idea that scientists normally uphold the interests of "science," and so scientific error will naturally self-correct, except in isolated cases of complaint.
  - Scientists presumed innocent until proven guilty.
  - Apt for research whose conduct or application has focused impacts, and those so impacted can easily identify themselves.
  - The agency is prohibited from raising inquiries of its own, when there are no complaints.
- Requires neutrality to the adversaries to establish pretext for adjudication.
  - Are the agency's interests independent of those of the adversaries?
  - Can the agency enforce sanctions based on its ruling (e.g. restitution)?
- Respected for its wisdom in past adjudications.
  - Judgments are proportional to the error, if any, committed.
  - Judgments deter similar cases from arising in the future.

the inquisitors are part cost accountant, part thought police. They ensure that the funders of scientific research are getting their money's worth by threatening to cut off the funding for scientific transgressors. Equally the inquisitors uphold scientific standards by threatening the transgressors with "excommunication," which would make it impossible for them to practice or publish as scientists.

In contrast, the accusatorial system assumes that scientists adequately regulate their own affairs through normal peer review procedures. Here scientific integrity is understood as a collective responsibility that is upheld by catching any errors before they cause substantial harm. The accusatorial system is then designed for those relatively rare cases when error slips through the peer review net and some harm results. This "harm" may involve concrete damage to health and the environment, or the corruption of later research that assumes the validity of fraudulent work. The legitimacy of the accusatorial system ultimately depends on the accuser establishing that some harm has been committed, which she alleges to be the fault of the accused.

In the Lomborg case, the CSD failed as both an accusatorial and an inquisitorial body. On the one hand, the CSD agreed to decide the case before an instance of harm had been clearly established. That certain scientists feel aggrieved that Lomborg's opinions carry more political clout than their own is not sufficient grounds for "harm," given that no adverse consequences were demonstrated to have followed from Lomborg's advice – minus a few bruised egos. The aggrieved would make better use of their time trying to defeat Lomborg-friendly politicians in the next general election. On the other hand, the CSD collected various opinions about *The Sceptical Environmentalist* without conducting a formal examination of those opinions. Thus, readers were left to speculate about how exactly this mass of opinion actually influenced the CSD's judgment. Were some sources trusted more than others? If so, why? Was Lomborg's scientific practice demonstrably worse than that of other environmental scientists? Is the problem here a particular researcher or an entire field fraught with mixed agendas? Clear answers to these questions would have enabled the CSD to demonstrate that it was operating with an independent standard of scientific integrity and not simply – as Lomborg's defenders claimed – allowing itself to be used to settle a political score.

Both the inquisitorial and accusatorial systems have much to recommend as models for a national research ethics board or an international court of scientific justice. Nevertheless, their complementary virtues do not sit easily together: should scientists be presumed guilty until proven innocent or vice versa? The two systems presuppose contrasting benchmarks of scientific propriety. For example, an inquisitorial system might set a standard that is above what most scientists are presumed to achieve, which would justify local "spot checks" on laboratory practice. In contrast, an accusatorial system might take a more *laissez-faire* approach, allowing the scientists themselves to set their own standards and perhaps even identify the wrongdoers for prosecution. The former would be tantamount to a police force in its operations, whereas the latter would invite subtler forms of self-regulation. I summarize the complications involved in trying to decide between the two regimes of epistemic justice in Box 5.3.

The question of judicial regimes for science is further complicated by the amount of responsibility that consumers of scientific research should bear in their dealings with scientists. Not only do scientists

---

Box 5.3 Complicating factors in assessing epistemic justice

- What is the benchmark of scientific propriety?
  - The inquisitorial approach sets a standard that may be above what most scientists practice.
  - The accusatorial approach accepts the *de facto* standard of scientists in a given field.
- If we apply these two benchmarks to "environmental science," where many – if not most – of the prominent scientists have clear political agendas, then we have the following two alternatives.
  - A "Grand Inquisitor" would establish his own benchmark of scientific propriety, potentially prosecuting **all** environmental scientists, regardless of their specific political stance.
  - The judge would wait for someone to actually claim that their own livelihood has been adversely affected by a particular scientist's work, since normally the field appears to operate with "permeable" standards.
- Why "dishonesty" might not be such a good category for thinking about scientific propriety, except in narrow cases like plagiarism.
  - The contexts in which scientists demonstrate relevance and validity in their own fields only partially overlap with those appropriate for policymaking.
  - The scientific research frontier is defined precisely by findings that can be taken in several alternative directions, which turn "overstated" conclusions into prescriptions.

commit errors in the theories they endorse, but policymakers and the public more generally also commit errors in the scientists they endorse. Moreover, both are entitled to make their own mistakes and learn from them – in ways that enable them to do more of the same in the future. This is what I call *the right to be wrong*, the first article in any decent Bill of Epistemic Rights (Fuller 2000a: ch. 1). The liberal professions of law and medicine have traditionally enforced this right by licensing practitioners whose competence is presumed until a charge of malpractice is formally lodged. This structure presupposes that a clear line can be drawn between the practitioner's expertise and the client's interests, on the basis of which one can judge whether the former has served the latter. To be sure, this line is increasingly blurred. But, arguably, the line is even *more*, not less, blurred in the case of strictly scientific expertise.

The public image of scientists as detached and cautious experts is not endemic to the scientific enterprise itself but *merely* to its public image. As Karl Popper saw very clearly, scientists *qua* scientists advance the course of inquiry by overstating their knowledge claims (aka "going beyond the data") in settings where they will be subject to stiff cross-examination and possibly falsified. In this respect, the entire issue of "scientific dishonesty" is misconceived, since scientists do not really need to believe what they put forward in the spirit of "hypothesis." Moreover, the drive toward overstatement may well be motivated by ideological considerations, as is clear in the case of environmental science. However, none of this need be a problem, so long as rigorous checks are in place. But how rigorous is rigorous? It is here that the public needs to take some responsibility for the conduct of science. However, the negative example of the Danish CSD shows that the answer does not lie in the restriction of debate.

My proposal moves in a contrary direction. A court of scientific justice – a more suitable title for the successor body to the CSD – should be empowered to determine whether the distribution of opinions in the media on a science-based topic is "fair" to the various scientific researchers and interest groups, as well as to members of the public who are not obvious "stakeholders." (This would be the court's inquisitorial side.) If the distribution is deemed unfair, then the court is empowered to redress the balance by releasing funds for the support of research and publicity into the underrepresented viewpoints. (This would be the court's accusatorial side.) To the maximum extent possible, these funds will be drawn from an independent source, so as *not* to involve a redistribution of the support already enjoyed by other viewpoints. This would help insulate the court from conflicts of interest and trade-offs that the public should make for itself. It would also underscore a basic insight of mass communications, namely, that a *relative* increase in the visibility of an alternative viewpoint is often sufficient to shift public opinion. The great advantage of the proposed court is that the import of its rulings would be, at once, to check and encourage scientific inquiry. Whether – and how – such a court would have ruled in the Lomborg case is, of course, speculative. Perhaps, given the media attention attracted by all sides to the case, it would have done nothing at all.

3. What Makes for Good Research: Good Character or Good Environment?

3.1. The case for environment

I recently found myself on a panel with distinguished British scientists and science journalists who appeared in agreement that the rise in research fraud among scientists was due to their lack of training in ethics. Without wishing to deny moral philosophers gainful employment, this diagnosis has the *prima facie* plausibility of blaming a rise in adultery on poor Sunday school attendance. The presumably upright scientists of the past never attended an appropriate ethics course, just as most loyal spouses failed to master Christian doctrine. This suggests that the source of the perceived change in behavior may lie in two other places: relatively new pressure placed on the scientists themselves and/or on those observing their behavior.

On the one hand, it may be that scientists commit fraud more often now because of added political and financial pressure to solve a problem or reach the truth quickly. On the other hand, it may be that such pressures are experienced less by the scientists themselves than by those responsible for exercising oversight over scientists. As the inquisitors scrutinize scientists' behavior more intensively, the more evidence of fraud they seem to find. Of course, both sorts of pressure may be operative, and they may feed off each other, as perhaps in the case of Professor Hwang, discussed earlier in this chapter. In any case, one is left with the impression that research ethics serves to scapegoat individuals for a systemic problem in the normative structure of contemporary science. In the previous section, I developed this point with respect to the Lomborg Affair. The lesson for the Hwang case would seem to be that we should reserve any harsh judgment on Hwang's conduct until its representativeness can be ascertained, which in principle would mean examining (perhaps randomly) the research of comparable labs.

From a strictly sociological standpoint, the emergence of research ethics as an area of public concern and a field of inquiry is a mixed blessing. If we judge its prospects in terms of the precedent set by medical ethics, then eventually it will be possible to justify any research practice *post facto*, thereby making the occurrence of misconduct harder to detect. This apparent decline in fraud would simply reflect the ability of the teachers and the examiners of research ethics to coerce researchers to abide by their jointly negotiated codes of conduct.

Moreover, research ethics may prove a big distraction. It effectively diverts attention from what transpires both before and after the actual conduct of research, which together constitute the true source of its value. It ignores, on the one hand, the large amounts of prior investment needed to do research, be it in training researchers, purchasing equipment, or funding grants. On the other hand, it also ignores the uptake of the research, again measured in multiple ways via commercial benefits, scientific recognition, career advancement, and so on. Instead, one becomes preoccupied with the attributes of the people, things, and processes "on site." Precisely because these attributes are viewed within such a restricted horizon, they are easily treated as intrinsic to the evaluated entities. Thus, typically lacking in the pursuit of research ethics is the charity implied in the 16th-century prayer said by heretics who were spared burning at the stake: "There but for the Grace of God go I." After all, given the competitive environment of contemporary scientific research and the big profits that await successful researchers, it is difficult to believe that those who happen to get caught are all who are potentially guilty of what remains a vaguely specified code of research conduct.

3.2. The case for character

Nevertheless, rather than casting a broad sociological net to examine the conditions that encourage researchers to cut corners and overstate their knowledge claims, research ethics has acquired an ideological superstructure known as "virtue epistemology," which harks back to classical Greek ideas that to be a reliable knower, one must be a person of good character (Zagzebski and Fairweather 2001). So, if one is shown to have been unreliable, he or she is punished for moral failure, and the research system is presumed to be intact – at least until the next culprit is convicted. Of course, virtue epistemology does not suppose that researchers are Cartesian agents existing in splendid isolation from each other. However, the normative glue that supposedly joins the researchers is an elusive force of mutual attraction called *trust*, which in practice simply interprets a liability as if it were a virtue. In other words, *ceteris paribus* (i.e. barring specific research ethics violations), the fact that we can't check our colleagues' work – either because we lack the time or skill – is taken to imply that we don't need to check.

An interesting defense of this position has been presented by the Anglo-American historian and philosopher of biology Michael Ruse who, on the basis of his book *The Darwinian Revolution* (Ruse 1979), became the founding philosophical expert witness in US trials relating to evolution and creationism. Ruse (2005) reprises signature cases from the history of modern biology, ranging from Darwin and Mendel to Theodosius Dobzhansky and E. O. Wilson, that arguably involved violations of trust: trust in authorship, competence, authenticity, originality, and so on. The most striking case is that of Dobzhansky, arguably the principal architect of the neo-Darwinian synthesis. Dobzhansky's success supposedly lay in his training in the two most polarized branches of biology required to forge the synthesis, natural history and experimental genetics, the former in the Soviet Union, the latter in the leading early 20th-century genetics lab, Thomas Hunt Morgan's at Columbia University. These two strands were eventually brought together in the highly influential *Genetics and the Origin of Species* (Dobzhansky 1937). Ruse shows, however, that Dobzhansky was more a jack than a master of trades who appropriated Sewall Wright's population reasoning without much acknowledgment or even comprehension. Dobzhansky's tireless data gathering was also sloppy and unreliable. Moreover, all of this was an open secret among biologists. Nevertheless, Dobzhansky passed without shame or rebuke because he was seen as having forged a whole much greater than its sub-optimal parts.

Ruse refuses to pass judgment on Dobzhansky. He is more concerned with the general question of why alleged trust violations in science are not prosecuted more vigorously. Because such allegations are usually aimed at very accomplished practitioners, argues Ruse, scientists suspect the settling of scores and hence question the motives of the accusers. But Ruse also argues that trust is so sacred in science that the very thought of its violation conjures up a sense of inhumanity comparable to the disgust felt about sexual perversion. Behind this melodramatic diagnosis lurks a sensible observation: the discovery of any kind of fraud offends the scientist's sense of fair play. But is any specifically *epistemic* damage done by scientists misrepresenting their activities? At this point it is worth recalling that Ruse is a scientific amateur. He founded and edited for many years *Biology and Philosophy*, the main journal in what is now the biggest growth area in the philosophy of science. Ruse accomplished all this without any formal training in biology. While this all-too-rare triumph of amateurism is to be applauded, as an outsider he may be too easily impressed by the allegedly "self-regulating" nature of the scientific enterprise.

There are two general reasons why we might be justified in "trusting" our colleagues in the purely behavioral sense of not checking whether they have done what they claim to have done: either our colleagues usually represent themselves correctly (and when they don't, they are eventually caught) or usually it doesn't matter whether they represent themselves correctly (and when it does matter, they are eventually caught). The latter possibility should be taken more seriously, at least from an epistemological standpoint. The more popular former option amounts to sociology by means of wishful thinking, i.e. a prior belief in collegial virtue discourages any further scrutiny. In a religious context, this would amount to unwarranted superstition. Elsewhere I have described this superstitious attachment to trust as an epistemic concept "phlogistemology," after the pseudo-element phlogiston, whose existence largely rested on negative evidence (Fuller 1996a).

4. Conclusion: Is There Justice in Fraud? A Long View

The metaphysically deepest reason why the misrepresentation of knowledge claims might not matter is that reality is more coarsely grained than our ability to represent it. The social constructions generated by our conceptual distinctions, from which divergent trails of consequences flow, are only as good as the system of reward and sanction designed to uphold them. We draw many true/false distinctions to which non-human reality is indifferent and so, if we do not check or at least provide the means for checking, reality is happy to live with whatever we take to be "true" or "false." Epistemologically speaking, this may be seen as a realist way of capturing the antirealist's position: most of the decisions by which we demonstrate our "reality principle" would make no difference to all other species and perhaps even most other human cultures. I happen to hold this antirealist position, but even full-blooded realists should appreciate why the preoccupation with research fraud might be overblown. Those guilty of such fraud have often reached substantially the right results but by empirically devious means, an "ends justifies the means" epistemology that in a more polite philosophical time had been called "intellectual intuition."

For example, Ruse suggests that Mendel's laws of heredity may have been derived from pea plant experiments too good to be true. This concern, originally raised by the statistician Ronald Fisher, is now attributed to overzealous efforts at "curve-fitting," whereby outlying data points were discarded to produce the simplest mathematical formulation of a set of empirical findings. Our willingness to excuse Mendel – indeed, to accord him the title "father of genetics" – rests on the subsequent history of biology, which vindicated the "laws" without his involvement (Brannigan 1981). Mendel's scientific salvation resembles Galileo's fate, at least on Paul Feyerabend's (1975) notorious account of a rather similar malfeasance with respect to the laws of physical motion.

There are two ways to think about Galileo's and Mendel's shared predicament: on the one hand, they may be seen as Roman Catholic heretics who took the Church's "double-truth doctrine" into their own hands. In other words, they provided the intellectual authorities with a pedagogical fiction (a "demonstration," in its original sense in geometry and physics) rather than their actual findings, as the Church itself had done for so many generations. What Galileo and Mendel presented as an appeal to open-mindedness when examining the data – not simply to be swayed by the prejudices of established authorities – could also be interpreted as catering to the reader's vanity and gullibility.

On the other hand, Galileo and Mendel can be seen as having taken epistemological realism deadly seriously, namely, that scientific methods are only as good as the discoveries they enable us to make. In that case, any lingering resentment of either visionary scientist merely reflects a preference to have science get to the truth by the epistemically approved means than to get to the truth at all, which may involve an intuitive leap beyond the phenomena. This conclusion was Feyerabend's way of rubbing the logical positivists' noses in their "methodolatry." Rather than try Galileo and Mendel for crimes against the intellect, Feyerabend would put the scientific method itself on trial, where Galileo and Mendel could then testify against the value of any such method, especially when interpreted in excessively empiricist terms.

Even those who dare not follow Feyerabend's lead here are forced to confront a tension between *truth* and *method* in science. It is analogous to what political philosophers routinely face between *morality* and *law*. The best response to the Feyerabendian challenge is a genuinely civic republican one: adherence to the scientific method is the best overall strategy to enable the entire polity to reach the truth in a timely fashion – that is, without too few running ahead of too many, which could easily create the conditions for either dominance from above or revolt from below. In effect, the scientific method amounts to an epistemic welfare policy that redistributes advantage from the quick- to the dull-witted to allow the most people to make the most use of the most knowledge. This staggering of the pace of scientific innovation is out of respect for our common humanity, which prevails over particular individuals' desire to jump ahead of the pack. Such a policy deserves the name "epistemic justice." To be clear, "humanity" in this sense is defined in primarily epistemic, not moral, terms: i.e. the capacity of every human to understand and, under the right circumstances, to have made a given scientific discovery. Thus, the appeal to method is designed to undercut the significance accorded to priority and originality, the two legal bases for lodging intellectual property claims in our times.

In any case, it is fair to say that Galileo and Mendel are not normally classified as perpetrators of scientific fraud. Indeed, they remain icons of scientific heroism, since in their own day both had to withstand dismissals for being seriously misguided, though perhaps not deceptive, in their scientific practice. A key difficulty facing a historical revisionist eager to hold Galileo and Mendel accountable to today's standards of research ethics is that so many subsequent scientists have built fruitfully on their allegedly fraudulent work. This would seem to justify the earlier claim that our two scientific heroes possessed intellectual intuition. At the very least, it points to the role played by *timing*, what the Greek sophists called *kairos*, in the revelation of fraud.

The longer it takes to reveal an alleged fraud, the harder for the revelation to have a moral bite, since by the time the fraud is revealed it will have probably opened up avenues of research that otherwise would have remained closed or delayed. While perhaps of little consolation to duped researchers, from a world-historic perspective, the opportunity costs of having been duped by the fraud would have been recouped in the long term by the valid research that ended up being built on what now turns out to have been invalid foundations. Perhaps the issue of fraud would hang more ominously over Galileo and Mendel today if the costs of believing them, both real and opportunity, were measurable and high. However, no money changed hands and no one's life was jeopardized in their cases. Indeed, there are no posthumously awarded Nobel Prizes – that is, highly desirable but scarce rewards – which would require that Galileo's and Mendel's own achievements (not simply the achievements they inspired in others) be directly compared with those of other potential recipients.

In this respect, the periodic reanalysis, and sometimes interpretive revision, of major scientific breakthroughs by historians seriously interferes with attempts to enforce strict codes of research ethics. After all, long undetected frauds have inspired normatively appropriate research, which suggests that the perpetrators pointed to a destination at which others eventually arrived safely. *Strictly speaking, then, those who commit such fraud are guilty more of confusing the potential and the actual than the true and the false.*

Indeed, had the history of science been populated by fraud busters of the sort countenanced today, it would be impossible now to distinguish between what philosophers call "realist" and "instrumentalist" approaches to science (Fuller 2000b: ch. 2). The former – exemplified by Galileo and Mendel (as well as Newton) – made claims about causal mechanisms that went well beyond the available empirical data. However, given the relatively decentralized and risk-free nature of scientific accountability, one could "speculatively" (which should be understood as a euphemism for "rhetorically" and even "deceptively") overshoot the commonly agreed body of evidence as long as it eventuated in fruitful research. In other words, realists have usually existed – with the striking exception of Galileo – as tolerated but skeptically regarded inquirers. It was only after W. K. Clifford's self-consciously anti-theistic address, "The Ethics of Belief" (1876), and science's corresponding professionalization, that responsible inquiry came to be defined in terms of demonstrating a sense of proportionality of belief to evidence.

In conclusion, the push to publish undoubtedly leads to fraud in the science system, but so what? An increasingly competitive research environment provides greater incentives to anticipate the results of research not yet done or to massage the data of results already in hand. But it equally provides more incentives to check for such transgressions of scientific propriety. Consequently, it is hard to say that there is now more fraud than in some supposedly less competitive past. We might imagine the level of fraud to have been less in the past, given the lack of incentives. But there may have been more fraud, given the lack of check. In any case, there are no records. What most certainly does *not* follow is that the relative failure to detect fraud in the past means that less fraud occurred. Just as much, or even more fraud, may have been committed in the past, but more rides on science today than ever before. Arguably, that is the real problem.

The current expression of concern about research fraud hides more systemic problems with the scientific enterprise. Interest in fraud is typically limited to the misrepresentation of research outputs, not the inputs. What is missing is captured by the phrase "citation clubs," whereby a circle of researchers cite each other's work to ensure publication in key "high impact" journals, regardless of their actual contribution to the intellectual basis of the research reported. As citations are increasingly used both to inform and to assess research performance, a subtly misleading picture is thus presented of the relative significance of particular researchers and their fields. Still subtler is the obverse phenomenon of "citation ostracisms," whereby rival researchers are strategically excluded so as either not to share credit for a finding or, at least as likely, confront criticism that might put the finding in a less glowing light. As with ostracism in past societies, the long-term effect of this sort of input-based fraud is to allow the ostracized researcher's career to die a slow and painful death – assuming she cannot find a new home elsewhere in cognitive space.

Moreover, certain kinds of fraud might actually be desirable, especially given science's tendency to quickly disown its past. Many plagiarism cases involve resurrecting work that was undervalued when first published but would unlikely appear credible now were it revealed to have been written many years earlier. In any case, much credible research can be – and has been – built on the back of frauds. Once that happens, the revelation of fraud may be reduced to a mere historical curiosity, as we have seen in the cases of Galileo and Mendel. Truth be told, science may flourish with a fair level of fraud because reality is more tolerant of our representations of it than we might like to think. *It may be that fraud safely goes undetected because it is "wrong" only in misrepresenting one's own work, but not in misrepresenting how reality works.*

No doubt readers will wonder whether my somewhat perverse reading of the history of science simply ends up encouraging the unscrupulous to enter science in the future. I do not believe that this is likely and, in any case, it does not matter, as long as the scientific establishment maintains its high tariffs on intellectual trade. The would-be fraudster would have to train for so many years to reach a level of competence – and an academic position – where she could get away with research fraud that she would probably deem careers in politics and business better prospects, in terms of effort vis-à-vis profit.

This suggests that the scientists caught in fraud cases are either normal scientists placed under undue external pressure to publish prematurely or aspiring scientific revolutionaries who felt they could idealize from the actual data, as Galileo and Mendel probably did – or perhaps a bit of both. In any case, they cannot be easily consigned to the ranks of the depraved.

Part III

The Transformation Problem

6

The Future of Humanity

1. Technology as a Religious Extension of Humanity

A striking but relatively unremarked feature of the "technology" side of "science and technology studies" concerns the strong Christian roots of its scholarship. This can be given a positive or negative spin. The canonical history of the social studies of technology was written by a Jesuit (Staudenmaier 1985), while the tale of "man's" (i.e. male-gendered humanity's) arrogation of the Earth, based on biblical privilege, is chronicled in Noble (1997). The most extreme sophisticated statements of the technophobe and technophile perspectives of the past half-century can be found in the work of two avowedly Christian thinkers: the French Protestant Jacques Ellul and the Anglo-Canadian Catholic Marshall McLuhan, respectively. Even from the heartland of STS, not normally known for its religious profundity, the field's two gurus, both Catholics, Bruno Latour and Donna Haraway, have adopted what Latour calls an "iconophile" perspective toward artifacts that would have done McLuhan proud (Latour 1999: ch. 9).

Implied in this fascination with technology are two contrasting visions, each turning on a sense in which technology is supposed to "enhance" the human condition. The first vision is of technology as the dematerialization, and perhaps even spiritualization, of the human condition, which aims to shift our ontological centre of gravity from animals to the deity itself. The second is of technology as an extension of human embodiment – "cyborganization," if you will – that enables us to become more integrated into a material world inhabited by other creatures. If the former vision stresses technology's capacity for translating and sublimating human ends into better means (i.e. "vertically"), the latter emphasizes our networking and interfacing with non-human things (i.e. "horizontally"). Yet both visions are opposed by self-declared Luddites like Ellul and Noble, who see here only the submission to false gods of convenience and an uncritical acceptance of automation. This chapter begins by explaining how science has helped to "sanctify" technology into a vehicle for human enhancement, and then proceeds to explore the various intended and unintended consequences of that development, one which largely coincides with the Westernization of the globe.

1.1. Science as sanctifier of technology

Technology pervades the human condition as a prosthetic extension of the physical infrastructure in which our biological evolution occurs (Basalla 1988). But it is neither unique to humans nor equally distributed across our species. Richard Dawkins (1982) has aptly referred to technology as a key component of the "extended phenotype" common to most animal species, as they transform their environments in ways that increase their chances for survival. And, as Darwin would have expected, the technological innovations made by *Homo sapiens* have reflected our species' ecological diversity, which should not be confused with a modern sense of distributive justice (Fleischacker 2004). For the latter, one must turn to the ongoing attempt to bring all patterns of human adaptation in line with the unified conception of reality associated with *science*, which from a world-historic perspective has been the acceptable face of "Western imperialism," namely, the systematic imposition of common standards for explaining, evaluating, and extending technology. Only once such uniformity was presumed, at least as a matter of principle if not fact, could one raise questions about the "equitable" distribution of goods, services, and opportunities beyond the spontaneous ("natural") variability of the human condition.

It follows that whatever universality science enjoys is more an achievement than an inheritance of humanity, something inscribed more firmly in international law and the college curriculum than the genetic code. Indeed, the genetic code itself came under the rule of science with the discovery of DNA in 1953, and has now become integrated into neo-imperial (or neo-liberal) intellectual property regimes.

Put as an analogy, the science relates to technology in intellectual history as the state to the market in political economy. From the standpoint of the history of technology, a market is nothing but an ecology domesticated by human concerns. Left to the market's devices, technology has tended to perpetuate, if not exacerbate, existing divisions between and within nations. This point is most obvious today from persistent asymmetries in the international production and the consumption of goods, contrary to what a free-trade view of globalization would suggest.

Consider communication technologies. Television, the paradigmatic passive good, spread into homes around the world much faster than an exemplary active good, the telephone – even though the latter was invented a half-century before the former (i.e. roughly 1875 vs 1925). At the level of production, televisions and telephones also interestingly differ in that the former were originally designed to collectivize and the latter to individualize users. These design features are not intrinsic to the technologies themselves, but rather served the interests of the original producers by creating what economists call "path dependence" in subsequent technological development (Page 2006). The resulting overall pattern is that the mass consumption of these artifacts amounts to a self-administered regime of social control, what Langdon Winner (1977) originally called (without need for Foucault) "autonomous technology."

Even more deep-seated knowledge-based asymmetries have been operative in the history of technology. These relate specifically to the alphabetic character of first telegraphy and then the Internet, which for the last 150 years has communicatively disadvantaged the West's main economic competitor, the Far East, which has relied on ideographic script (Standage 1998). To be sure, these disadvantages have been eventually overcome through still more technological innovation. For example, as of this writing, there are more weblogs in Japanese than in English. Nevertheless, the very fact that the Far East was forced to "catch up" highlights the role of path dependence in the history of technology.

Seen from the standpoint of the history of technology, science appears paradoxical: while retaining the marks of its origins in a secularized Protestant Christianity, science has managed to supplant local forms of epistemic justification throughout the world. In effect, science displays its own kind of path dependence. Specifically, science has aimed to realize the universalist dream of drawing the monotheistic religions and the Enlightenment ideologies into a single narrative that heads for a regime in which humans dominate nature without dominating each other. Science's solution, very much in the spirit of democracy, is that whatever coercion is exerted over humanity is exerted equally over all of its members. This is what Karl Popper (1945) called the "open society," an ideal that we have already explored in these pages. In practice, it has meant that all scientists – and, in a rational polity, all citizens – must submit to the strictures of the "scientific method," which is popularly associated with "logic," "evidence," "reason," or, more usually, some institutionally sanctioned combination of practices associated with these terms.

Consider something as taken for granted as the ease with which scientific discussions shift out of native languages to a technical version of English in both writing and even speech. The religious roots of this practice are pertinent in a way that typically elude STS researchers infatuated with science's "hybridity": science never quite blends in with local cultures as simply one among many ingredients. Its presumptive superiority as a form of knowledge is always present, if only as linguistic alienation, where English now functions as Church Latin once did, a remnant of the exclusivity of faith demanded by the Abrahamic religions. As failed theologians from Hegel to Heidegger have realized, science's requirement that all knowledge claims be subjected to a common standard of "truth" is simply a depersonalized – "onto-theological," as Heidegger would say – version of the equal subordination of all humans to The One True God. Otherwise, technology's house philosophy, an empiricism relativized to life circumstances, would be tolerable, at least epistemically, if not ethically (i.e. from the standpoint of distributive justice).

Science reveals its Abrahamic descent in its refusal to ground authority in such biologically salient features of the human condition as one's parents, physical strength, or personal attractiveness. Indeed, from a strict Darwinian standpoint, to be subject to the rule of science is to systematically distort, if not pervert, natural selection – at least insofar as science projects a sense of objectivity that transcends the specific environments in which humans (and other organisms) are required to survive. Indeed, science provides the sharpest example of the distinction between cultural and natural selection, since the people who typically most benefit from science are those whose capacities and interests are very different from, if not opposed to, the scientists themselves. In the case of medical research, these would include, on the one hand, the physically and mentally infirm and, on the other, the financially motivated (assuming scientists are primarily truth-seekers). However, this twist in the Darwinian tale only became apparent once science was harnessed by industry to produce technologies on a potentially global scale – ironically around the time Darwin's own theory was being crafted (Marks 1983: 186–221; Inkster 1991: ch. 4). In other words, the very same scientific mentality that enabled Darwin to arrive at the theory of evolution by natural selection also enabled others to obviate the effects of natural selection, a point that "Darwin's bulldog," Thomas Henry Huxley, astutely observed (Fuller 2006b: ch. 11).

1.2. Science as technology on a global mission

The idea that technology is the appliance of science derives from the West's historical deployment of science to expand the European sphere of influence overseas. Science was a "technology of control" in several senses (Beniger 1986). When the natives appeared at least as morally literate and technically sophisticated as the Westerners (i.e. "civilized"), science exerted control by encouraging sustained trade with the natives. However, when the natives were deemed less sophisticated, science was applied to gather data, samples, and reorganize – and sometimes destroy – native socio-economic structures in order to render native regions governable by Western standards. In this context, colonial science was an extension of colonial administration, whose reward structure and sense of purpose remained heavily indebted to the "mother country" (Basalla 1967; Pyenson and Sheets-Pyenson 1999). A continuing legacy of this relationship is the "brain drain" from the former British and French Empires to the UK and France. From a global standpoint, the spread of Western science resulted in a division of intellectual labor whereby the Americas, Africa, and Asia provided raw "human capital" that was processed in – and often retained by – mother country institutions.

It is worth observing that the path of globalization forged by science-based technology created high expectations among the colonials that arguably set them up for eventual disappointment. Thus, what Andre Gunder Frank called "dependency theory" in the 1960s was a theoretical expression of Latin American resentment to the political and economic impediments that prevented their countries from reproducing the path taken by their European parents and then North American cousins, to become equal trade partners in the global marketplace (Blomstrøm and Hettne 1984). This sense of unfulfilled promise relates to the introduction of experimental natural science into higher education, combined with state-sponsored university-industry collaborations, which tightly bound together the intellectual development of the mother country and its colonies in networks of technological innovation. Late 19th-century figures like Kelvin in the UK and Siemens in Germany exemplified the staking of national reputations on global technoscientific prowess, a practice that continues to this day through the rhetoric of "global competitiveness." By the first decade of the 20th century, they had informed the first genuine model of global governance, *imperialism*, which Lenin soon converted into the template for World Communism.

However, this vision of technoscientific globalization came to a crashing, albeit temporary, halt in 1918 with the defeat of the world's leading scientific power, Germany, in the bloodiest war the world had yet seen. The aftermath of World War I thus witnessed a strong backlash against the marriage of science and technology. Which of the two partners was to blame remained an open question. On the one hand, positivists and Marxists were keen to protect science as a progressive social force from its adverse technological applications. They stressed the "value-neutral" character of science itself, as opposed to its "value-laden" application as technology. On the other hand, phenomenologists and regionalists revived an interest in craft technologies whose local cultural bases do not rely on a globally hegemonic science for legitimation. In this context, Oswald Spengler's popular *The Decline of the West* and Martin Heidegger's more esoteric philosophical works faulted the West's specifically "Faustian" desire for knowledge that, if it is to be had at all, is best left to God (Herf 1984).

After the scientific application of technology reached new heights of sophistication and destruction in World War II, a new generation of protests was unleashed against what by 1960 was called the "military-industrial complex," a phrase coined in US President Eisenhower's farewell address. Generally speaking, the positivists and Marxists on both sides of the ongoing Cold War lived a schizoid existence, conforming to the norms of both "basic research" (another coinage of the period) and national security, which demanded different outputs, one associated with "science," and the other "technology." The phenomenologists and regionalists became devotees of natural ecology, which typically entailed a "small is beautiful" ethic on a global scale that would subvert the legitimacy of large nation-states with imperial ambitions. However, as the Cold War came to a close with the fall of the Soviet Union in 1989, allegiances were scrambled by an ascendant neo-liberal logic of "globalization," in which it became possible via "corporate environmentalism" for transnational firms to support a green agenda by introducing "clean technologies" into their production processes in a way that eludes the labor laws of declining nation-states (Hoffman 1997).

A counter-current to this overall line of thought had already emerged in the mid-19th century, once a generation learned of the costs and benefits of various strategies of industrialization. Germany, the United States, and Japan successively capitalized on what the economic historian Alexander Gerschenkron (1962) called the "relative advantage of backwardness." These three latecomer nations did not simply copy the technologies of the leading nations, but designed their own versions capable of satisfying comparable needs more efficiently, which were themselves subsequently subject to foreign uptake (Inkster 1991: ch. 5).

Prior to Gerschenkron's critique, which was inspired by Schumpeter (1950) on entrepreneurship, cross-cultural learning was presumed to be a matter of imitation, with the recipient culture often seen as producing pale imitations of the original artifact. Otherwise, the culture was presumed to have invented its technologies *de novo*. Emile Durkheim's rival for the establishment of sociology in France, Gabriel Tarde, was the historic champion of this older bipolar theory of invention and imitation, which continues to travel under the term "diffusion."

The above discussion suggests that the globalization of technology can be explained by three mechanisms, which can be defined in terms of what is transferred across locations: (1) *migration* (of people); (2) *diffusion* (of artifacts); and (3) *translation* (of function). These three possibilities may be treated as successive stages in the alienation of technology from expertise, which in turn serves to distribute knowledge more widely, at least in terms of reducing information asymmetries between innovators and adopters. Migration requires foreign experts to operate the technology, whereas diffusion presupposes the codification of instructions that are generally available to natives, and, finally, translation concerns the reproduction of the foreign technology's functions (e.g. through reverse engineering) in native media.

Accounts of technology transfer have historically reproduced this process, with migration quickly yielding to diffusion, and translation still relatively unexplored. Biblical echoes of a single historic origin to humanity remained in anthropological accounts of artifacts diffusing from a single origin as late as the 1890s, say, in the work of Friedrich Ratzel and W. H. R. Rivers (Harris 1968: ch. 14). They argued that oceanic migration explained similar clusterings of artifacts in different parts of the world. Diffusion was presented as more sensitive to the historical and ethnographic record than evolutionary strategies that assumed, often on the basis of vague cross-cultural analogies, that all cultures independently pursue similar paths of development. However, the anthropologist Franz Boas criticized diffusionists for denying inventiveness to all of humanity, since all artifacts would seem to have originated in the same set of places, which usually included Egypt. A stronger version of this critique simply denies that all cultures must arrive at the same artifacts, given their different environments. Here the diffusionist responds that this only reveals the role of outside factors in stimulating cultural change. That cultures can learn from each other is a better indicator of the "psychic unity" of humanity than parallel evolutionary paths. Whatever its empirical validity, diffusionism clearly legitimated colonial expansion during imperialism's heyday (Basalla 1967).

In the 1960s, diffusionism was revived as "technology transfer," with an emphasis on explaining why agricultural and medical innovations did not diffuse more quickly in rural settings, especially in the Third World. The problem supposedly lay in the reception of the innovations rather than the innovations themselves, with historical precedents drawn from the legal barriers to new inventions in feudal Europe and the Far East. US sociologist Everett Rogers (1962) modeled diffusion on mass communications, distinguishing between producer and consumer initiatives, i.e. the migration of technical personnel to train locals and the adaptation of new techniques to local needs. Economic historian Nathan Rosenberg (1972) fashioned a similar distinction to explain differences in the diffusion of industrial innovations between the US and the UK in the 19th century. The US pattern was producer-driven, leading to massive capitalization and labor replacement, high-volume manufacturing, and product standardization. The UK pattern was more consumer-driven, encouraging greater product diversification, with little standardization even in communication and transport technologies.

While leading to greater economic growth, the US model has proved to be a poor model for diffusion in the Third World, where labor remains the primary source of value, protectionist trade policies are relatively weak, and massive capital investment is always a risky option. Consequently, by the late 1970s, talk of technology transfer had begun to yield to talk of "appropriate technology," in which translation replaces diffusion as the mode of technological transmission from the First to Third Worlds. This move has corresponded to a general skepticism about the efficacy of top-down technology management strategies. As already highlighted in Latour (1987), this was the context in which the leading STS research program, actor-network theory, was first developed, which in turn reflected Latour's own doctoral research in the early 1970s on the failure of French management courses in the recently decolonized Ivory Coast (Fuller 2006c).

2. The Elusiveness of Technology's Human Touch

That technology might distinguish humans from other animals was most vividly put forward by Karl Marx in volume 1, chapter 7, of *Capital*, where he suggests that the human species should be reclassified as *Homo faber*. As we have seen, a strict Darwinist like Richard Dawkins would chastise Marx for his hasty anthropocentrism. However, Marx appears in a somewhat better light if we render *Homo faber* as "man the fabricator." The ambivalent "fabricator" improves on the more neutral "maker" as a translation of *faber* because it captures both the historical source of Marx's inspiration and the uncertain ends to which humanity's self-understanding might be put.

Marx's source was Benjamin Franklin, who notoriously regarded technology as an effective way for people to disguise their intentions by inventing things that even their enemies would find useful. Franklin, a believer in economy in all things (including the expression of one's motives), clearly saw the debt that modern technologies owed to the persuasive techniques of classical rhetoric: hidden costs are not questioned if the benefits are provided upfront, a lesson that successful marketing teaches on a daily basis. Thus, Franklin's response to the Darwinist would be that technology enables the artificer to control the environment in which selection occurs, so that the foresight of at least particular humans can replace the blindness of natural selection.

Franklin's perspective is comprehensible in terms of behaviorist psychology, according to which technology accomplishes an economy of thought by shortening the time between stimulus and response. What previously would have required considerable and deliberate effort is reduced to an automatically triggered experience or routine. Defenders of this approach in the modern era, notably Marxists and positivists, saw the value of technology in terms of the time it left over for "higher pursuits." It is as if technology would allow humans to transcend the drudgery in their lives and thereby escape their animal origins. They could then spend "quality time" thinking about quality things. This state, which was often likened to creating a "heaven on earth," has its roots in medieval Christian justifications for the pursuit of the mechanical arts as an analogical extension of the creative powers of the deity. The labor theory of value common to Aquinas, Locke, and Marx is a legacy of this line of thought. It is complemented by the idea that nature is "raw material" to which human effort alone can provide purpose and form (Noble 1997).

However, this progressive view of technological change has been opposed on two grounds related to an unintended consequence of the spread of industrial and, more recently, computer technology. They share a realization that the technological imperative of efficiency tends over time not to economize itself but to become more profligate. Together, the two grounds have provided science fiction with a wellspring for dystopic visions of a "post-human" future (Fukuyama 2000: ch. 1). In the first case, humans can be seen as regressing to service providers for their animal passions. In the second case, they appear as passionless robots of pure reason.

2.1. Humans unintentionally become animals

The first unintended consequence is that the distinction between cognition and sensation, which traditionally elevated humans above other animals, is erased, as the former is reduced to the latter. In recent times this critique has been lodged in terms of the globalization of the consumer mentality, the response to which Herbert Marcuse (1964: ch. 3) famously diagnosed as "repressive desublimation." In other words, a potentially unruly populace is pacified as an array of appetites is regularly activated and sated. Aldous Huxley's 1932 dystopian novel *Brave New World* anticipated this state of affairs, whereby humans are effectively reduced to patients dependent on the all-purpose food-drug "soma."

However, technology's tendency to reduce thought to automatic responses applies at least as strongly to ancient empires – including the Egyptian, Indian, Chinese, Roman, Islamic, and Incan – whose political infrastructure was dominated by massive civil engineering projects of often incredible ingenuity that aimed to maximize harmony by channeling the movements of the various peoples under their rule. This use of technology as a naked exercise of power was probably much stronger then than now, if only because the knowledge of the principles underlying the technology – and a corresponding culture of scientific criticism – was much more restricted.

In light of this historical pedigree, Hegel famously dubbed this conception of technology "Oriental Despotism," which Marx, and after him Karl Wittfogel, dubbed the "Asiatic mode of production" (Dorn 1991: ch. 1). Technology here provided the material infrastructure for the earliest attempts at global governance by intermediating vast regions of great cultural diversity in networks of roads and other public works, especially irrigation projects. The historic exemplar of this kind of regime was China – indeed, until at least the late 18th century, when it was finally challenged as the world's economic superpower by the combined forces of Europe and its American colonies. Characteristic of this infrastructural use of technology was that the dominant power taxed its subject-regions but left their local modes of production and social relations largely intact. Technology was promoted as a stabilizing force that constrained the channels of communication and transportation.

This was quite unlike the creative or innovative function of science-led technology in the Europeanized world, where local economies were radically restructured ("rationalized") to make them efficient producers of surplus value for constantly shifting and expanding markets. World-systems theorists following the lead of Immanuel Wallerstein and especially Andre Gunder Frank have interestingly attempted to reverse the progressive image of Europe vis-à-vis Asia on this score by casting the West's drive toward technological innovation as systematic piracy, designed to blackmail the rest of the world into revaluing its goods and services. This original "gunboat" version of globalization was epitomized in the mid-19th century by the stationing of British and American military might in Chinese and Japanese harbors. All of this was done in the name of "opening" markets, but it resulted in trade agreements that typically short-changed Asia's labor-intensive mode of production, leading (most noticeably in the Indian subcontinent) to a general decline in the standard of living – except in Japan, which quickly engaged in a policy of defensive modernization (Inkster 1991: ch. 9).

Gunder Frank (1998) stressed that as late as 1800 China still boasted larger and freer markets, more private property, a higher standard of living, and at least as productive use of land and labor as the UK. Moreover, Chinese economic success was propelled by an indigenous version of the capitalist ethos that stressed efficiency and the reinvestment of profits. But also like the UK at that time, China was suffering from diminishing returns from the land, the primary source of wealth in both societies. Soon thereafter, as China continued along the same trajectory, the UK "industrialized" its economy, which involved the radical reorganization of land and labor, alongside the substitution of goods and services through technological innovation, resulting in the literal manufacture of new sources of wealth and, more importantly, new "economic" ways of valuing wealth (Beniger 1986).

In retrospect, the most striking feature of Britain's transformation was the relative ease with which land and even labor were converted to raw material that could be formatted to purpose. The history of technology as a field of study in Britain was founded in the early 19th century on the emergence of concepts like "energy" and "efficiency" as both the ideological superstructure and scientific foundation of this transformation (Cardwell 1972). In contrast, the view that remained dominant in China – the need for humans to be in harmony with nature, understood in quasi-sacred terms – diminished and became a romanticized, if not reactionary, presence in liberal Britain (Polanyi 1944).

The difference in technological trajectories pursued by Britain and China is traceable to the ideological shift associated with the 17th-century Scientific Revolution, whereby the West started to take seriously the Abrahamic message that humans are created in "the image and likeness of God" and that God needs humans for creation to be complete. This view was already anticipated in the *kalam*, Islam's rational theology, which portrayed Allah as free to create the world from moment to moment, reproducing it usually but occasionally changing its structure. This vision of humanity as a micro-deity informed by divine open-endedness fueled technological innovation in medieval Islam, whose imperial reach was comparable to China's and similarly exhibited an "Asiatic mode of production." It was through trade with the Islamic world – coerced or otherwise – that Christendom managed to stage its Renaissance, starting in the 15th century, which eventuated in the Scientific Revolution (Frank 1998: ch. 2).

Where Christendom excelled over Islam was, ironically, in its failure to regulate social intercourse between laborers and scholars (Fuller 1997: ch. 5). Mostly this reflected Christendom's relatively disorganized character, but there was also a doctrinal basis for free exchange. Just as God requires what he creates to complete creation, so too humans require what they create to do likewise. Thus, the technologies associated with experimentation become both prosthetic devices to extend human cognitive capacities and standards against which those capacities are evaluated. The computer's dual role as the extension and the measure of rationality in the modern era is a clear case in point (Fuller 1993: ch. 3). By contrast, although China had much of the relevant technology and some of the theory, the two were pursued independently of each other. Human reasoning processes were thus treated as self-contained, or at least not improvable or revisable by technological mediation. Under such conditions, modern science could not develop, let alone feed back into the development of technology.

2.2. Humans unintentionally become artifacts

The second unintended consequence is subtly different: thought is not only reduced to an automatic response, but automation is itself elevated to the standard of cognitive achievement. One of the most concrete examples is the measurement of intelligence via IQ tests, which demand the generation of many quick and accurate answers to well-defined questions. However, this practice must be seen against a larger project to construct a "frictionless medium of thought" capable of taking the most efficient path to the truth. The ascetic quasi-spiritual character of this project is traceable to Descartes' mechanical theory of mind, which came to inform Taylorist theories of "scientific management" in the 20th century, as well as more specifically computer-based research in "artificial intelligence." By the early 19th century, it had already become a principle of social organization with Count Henri Saint-Simon, who applied the efficiencies generated by the capitalist division of labor in industry to society as a whole, presaging modern notions of bureaucracy and social engineering (Hayek 1952).

Crucial to Saint-Simon's explicitly "socialist" vision was the idea that a common body of information – what we now call "social statistics" – could be obtained from people to maximize the efficient administration of their wants and needs. Saint-Simon's follower, Auguste Comte, extended these efficiencies of functional differentiation to knowledge production itself. This was the context in which he coined "sociology." It is the source of modern arguments for the rational necessity of scientific specialization. Interestingly, Comte's sense of efficiency was founded on the Roman Catholic principle of "subsidiarity," whereby decision-making is, as much as possible, delegated downward to the point of impact. This strategy presupposed a "natural law" perspective whereby under normal circumstances ordinary people are best positioned to register their experience, though the best way to deal with it may require higher-order delegation. This systematic empiricist perspective, which Comte believed was a lesson that the history of science could teach contemporary society, is what became known as "positivism."

Sociology has always struggled with its status as both an empirical and a normative discipline, an ambiguity preserved in the meaning of "social engineering," which as Max Weber keenly realized, mixed empirical means and normative ends. While Comte envisaged sociology and positivism as the twin spawn of a political movement dedicated to the scientific reformation of society, national traditions have subsequently joined the struggle differently: the French proposed secular replacements for the Church as a basis for social solidarity, while Germans debated whether their research questions should be dictated by policy relevance and the Americans pondered the universalizability of their exceptional collective experience. Yet, in all these cases, it has been difficult to delineate the scope of sociology without taking seriously our biology as the social engineer's raw material, regardless of who issues the marching orders. As each of the great national traditions in sociology came to self-consciousness at the end of the 19th century, the struggle for defining the discipline was refracted through Darwin's theory of evolution by natural selection, a point that remains airbrushed out of even sophisticated histories of sociology (e.g. Levine 1995; cf. Fuller 2006b: chs 4–5).

3. Social Engineering, or Sociology's Hidden Agenda

Saint-Simon and Comte's equation of *socialism* = *sociology* = *social engineering* was indebted to a vision of technology as applied science that began to be persuasive in the 18th-century European Enlightenment, when science was thought to be close to completion, thanks to Newton's unified explanation of terrestrial and celestial motions. The main imperative then was to spread the Newtonian paradigm as widely as possible via "technology," understood in its widest sense as socially beneficial artifice, which included, say, the US national Constitution, founded on a separation of powers governed by checks and balances (Cohen 1995). In Comte's France, influenced by the spirit of Napoleon, social engineering was extended to the construction of carceral institutions like prisons, hospitals, and schools, as well as affiliated modes of transport. By the end of the 19th century, social engineering had begun to take a more invasive turn with the advent of mass-produced drugs dispensed in solid, liquid, and gaseous states, with Germany leading the way. Now, at the dawn of the 21st century, social engineering has the potential to go deeper under the human skin with nano-based biotechnology (Roco and Bainbridge 2002). A good sense of social engineering's engagement with human biology can be gained from an untold part of the history of sociology in the United Kingdom.

Social engineering is routinely omitted from the history of British sociology. In one sense, this is quite understandable. Many of the figures mentioned below – often equipped with natural science expertise and Marxist politics – presumed that sociology was epistemically much more advanced, and hence directly relevant to policy, an assumption no longer held by today's sociologists. This raises an interesting point: the two great theorists of 19th-century British liberalism, John Stuart Mill and Herbert Spencer, still understood "theory" in the Enlightenment sense of a social policy designed to be tested against the behavior of the body politic. In contrast, sociologists today see "theory" as operating at a level removed from ordinary social life, where theorists test their speculations against each other's opinions, all trying to divine sociology's fundamental principles. It is as if sociology has descended into a bleak medievalism in the wake of the failure of the Nazi and Soviet social engineering projects associated with World War II and the Cold War. In any case, for sociologists today (and not only in Britain), biology is the ultimate "other," understood as unified and undifferentiated – *sui generis*, as it were (Giddens 1976: 160; cf. Fuller 2006b: 81).

However much they tried to distinguish themselves from Comte, not to mention each other, Mill and Spencer accepted the basic premise of Comtean sociology – namely, that the principles of human nature and social change had been largely established, and that the remaining questions were ones of rational application and just administration. To be sure, these tasks required the acquisition of new knowledge, but of a more specific, "policy-relevant" sort, gathered on a "need-to-know" basis, certainly not for the purpose of establishing a self-perpetuating academic speciality. In effect, sociology was a kind of *macro-medicine*, whose therapeutic poles were defined by, on the one hand, letting nature run its course on the social organism (Spencer) and, on the other, promoting the body politic's native capacities (Mill).

Within the larger dynamic of British intellectual life, Mill and Spencer's sociologically informed battle for the soul of liberalism was joined in opposition to a conservatism dedicated to upholding cultural standards against their otherwise inevitable leveling in the face of democratic enlargement. Lepenies (1988) reviews the successive generations of this pivotal battle, which pitted Jeremy Bentham against Samuel Coleridge, T. H. Huxley against Matthew Arnold and, not least but least celebrated, Karl Mannheim against T. S. Eliot. The league of liberals united against such conservative foes is typically glossed as British sociology's historic "utilitarian," "empiricist," or "atheoretical" streak. However, this superficial reading overlooks Comte's deeper imprint on the national tradition, namely, his view that sociology integrates the sciences for the purpose of social engineering.

In this context, the expression "social engineering" should be understood as ranging from the "piecemeal" approach favored by Karl Popper, that is practiced in contemporary parliamentary politics, to the Soviet-style top-down version advocated by J. D. Bernal and demonized by Michael Polanyi and Friedrich Hayek. Unsurprisingly, the leading sociological émigrés from Nazi Germany, Otto Neurath and especially Karl Mannheim, easily inserted themselves into this discourse. It was even promoted as part of the BBC's commitment to public service broadcasting in the postwar period. Bertrand Russell addressed social engineering's prospects in the first set of Reith radio lectures in 1948. Echoes of his discourse survive – in a decidedly depoliticized and desociologized form – in the scholastic disputes between methodological individualists and holists (O'Neill 1973) and, especially after C. P. Snow's popular 1956 Rede Lecture, "the two cultures problem."

Notwithstanding the absence of the social engineering debate from British sociology's historical self-understanding, the social engineering projects associated with the rise of the welfare state, in which the UK played a seminal role, remain an enduring social scientific legacy (Duff 2005). Indeed, a coherent institutional and intellectual history of sociology in Britain could be told as the long-term pursuit of several quite different and often ideologically opposed social engineering strategies, ranging from the largely biological to the largely cultural, to overcome generations of inherited class differences.

The revival of the social engineering tradition would address, and possibly heal, the rift between researchers who profess, on the one hand, "social theory" or "empirical sociology" and, on the other, "social policy" or "social work." This rift remains inscribed in the names of UK academic departments. It amounts to institutionalized intellectual snobbery, occluding the latter pair's status as the former pair's disinherited siblings. Social policy and social work are the legatees of the social engineering tradition. What appears to social theorists and empirical sociologists as a lack of theoretical sophistication and intellectual curiosity on the part of policymakers and social workers is really the latter's residual social engineering impulse – that sociological knowledge is already sufficiently robust to be applied, albeit critically, in society at large.

Of specific relevance here is the biological strand of British social thought, often demonized as "eugenics," but nonetheless registered in William Beveridge's view of "social biology" as the foundational social science needed for administering the welfare state. Such a science gathers the social organism's "vital statistics" on the basis of which static and dynamic "quality of life" indicators (e.g. life expectancy and class mobility) are measured and, where necessary, corrected. The relevant modes of intervention, ranging from basic medical and educational provision to more genetically targeted treatments, can all be seen as strategies to alter the British population's selection environment with the aim of welfare maximization. Here it is worth recalling that the welfare state's policy objective of "equal opportunity" was coined by Julian Huxley, the embryologist who also coined the phrase "evolutionary synthesis" to characterize modern biology's marriage of Darwinian natural history and Mendelian genetics. For Huxley, the state's provision of "equal opportunity" enabled the conversion of society into a level playing field modeled on the laboratory, which allowed the full range of the population's genetic endowments to be expressed without the biasing environmental factors of the family and the local culture (Werskey 1988: 243).

That some non-human "Nature" might place a normative limit on human potential had attracted Spencer but repelled Mill in Darwin's original presentation of evolution. Moreover, before the evolutionary synthesis was canonized in the 1940s, geneticists and molecular biologists tended to regard the Darwinian fixation on the exact timing and placing of species in the earth's antiquity as unnecessary encumbrances on their forward-looking pursuits, much as psychologists and economists still largely treat history and anthropology. (The double-edged nature of the comparison is deliberate.) Julian Huxley, while giving Darwin his due as a natural historian, believed that the late 20th century might come to realize Lamarck's doctrine of the inheritance of acquired traits – but now under the aegis of human, rather than natural, selection.

3.1. Transhumanism and the advent of biotechnology

The spirit of Huxley's vision survives in another of his coinages, "transhumanism," which is nowadays associated with the technological enhancement and extension of the human condition known as *cyborgs*, the subject of the final section of this chapter. Even without becoming a convert to science fiction or postmodernism, one can appreciate the roots of this mentality in the British social engineering tradition. The relevant backstory would revisit Beveridge's choice of the zoologist and demographer Lancelot Hogben as the standard-bearer for "social biology" in his directorship of the London School of Economics (LSE), shortly before crafting the British welfare state (Dahrendorf 1995: 249–6). Hogben was a central figure in Bernal's scientific socialist circle, whose members – including Huxley – waged a remarkably popular campaign on behalf of social engineering throughout the middle third of the 20th century (Werskey 1988).

The securest link between this tradition and our own times is probably Conrad Waddington, Professor of Animal Genetics at Edinburgh University until his death in 1975. Waddington is interesting for various reasons that will become increasingly clear in the rest of this chapter. Not least of his achievements was to have proposed the first "social studies of science" program, the institutional basis of the so-called Edinburgh School in the Sociology of Scientific Knowledge, the prototype for a reflexively scientific study of scientific practice nowadays represented by STS, which would have met with Comte's approval, at least in principle if not in actual detail (Fuller 2000b: 327–8). Moreover, Waddington's final scholarly act was to pen the controversially positive review of E. O. Wilson's *Sociobiology: The New Synthesis* in *The New York Review of Books* that arguably launched the current wave of biologized social science (Segerstrale 2000: 18–24).

But three decades earlier, Waddington had already introduced the concept of "epigenetic landscapes," which was designed to capture a potentially progressive consequence of the inherent genetic variability in any population – namely, that some population members will probably have been pre-adapted to any new systemic changes undergone by the environment, which in turn would give them a selective advantage in the future that would shift the overall development of the species, perhaps even resulting in a new species. In other words, traits that previously had no use or were perhaps even mildly dysfunctional suddenly become valuable because the world has changed to make them so. This was Waddington's attempt to translate into Darwinian terms the Lamarckian idea that organisms somehow genetically incorporate their experience (Dickens 2000: ch. 6). The result would be biotechnology by ecological manipulation of the sort that might be converted into design features of housing and other features of the "built environment."

However, "biotechnology," a phrase that dates back to Weimar Germany, originally meant the merging of organic and mechanical elements in a common industrial process. In the 1920s it referred to the use of living organisms, or enzymes produced by them, for the manufacture of consumer goods like detergents and drugs. However, since the 1950s, the term has been reserved for applications stemming from recombinant DNA research, especially the transfer of genetic material from one organism to another, sometimes of a different species (i.e. "xenotransplantation"). Whereas the original sense of biotechnology strategically inserted organisms into a mechanical process, the latter sense uses mechanical means, "gene splicing," to enhance organic performance.

The success of both sorts of transfers, especially the latter, has been of considerable theoretical significance within biology (Smith 1998). A standard method for evidencing common descent among species of organisms, a basic tenet of modern evolutionary theory, is to show that genes taken from one species can trigger analogue effects when implanted in another species. Thus, a light-sensitive gene from a mouse can enable sight in a fruit fly, even though the structures of their eyes are radically different. Here neoclassical economists might recognize the conversion of functional equivalence to equality of value: call it "genetic fungibility." After all, the fact that the gene comes from the mouse does not seem to undermine the fly's ability to use it for seeing. More sinisterly, it may also constitute a more intensive version of what Marx feared would happen as technology replaced human labor as the source of value in industry. Both phases of the history of biotechnology have been productive of "cyborgs," or "cybernetic organisms," a phrase coined by the mathematician Norbert Wiener in 1960 for organisms designed to incorporate part of their environment in a self-regulating system (cf. Wiener 1948).

3.2. The cyborg moment: Bateson to Haraway

The evolution of cyborgs has proceeded in three stages: (a) technology is externally applied as a prosthesis that extends the organism's function, as in the case of eyeglasses to correct myopia; (b) technology is incorporated into the organism, as in corrective surgery to the lens and cornea of the eye; and (c) technology is introduced less obtrusively through "nano-machines" in the nervous system that regulate the organism's vision, compensating for any genetic deficits, so that the organism never becomes myopic. To be sure, this three-stage logic points to the role of process miniaturization in the erasure of the machine–organism boundary. But it would be a mistake to conclude that "cyborganization" always tends towards the microscopic. The original cyborgs envisaged by Wiener, prominent in Cold War military strategy, were global models of political economy whose legacy persists in, say, computer simulations of climate change. These were inspired by Keynesian macroeconomics, which features a central banker as "system governor" who thermostatically controls money supply and sets parameters on prices to ensure that the economy remains in "general equilibrium" through the perturbations of the business cycle (Mirowski 2002).

This vision became widespread in economics through the Cowles Commission, first convened in 1932 to prevent the occurrence of another Great Depression. But the devastation wrought by World War II provided an incentive to extend the vision to a global strategy for preventing a third world war, namely, by controlling the systemic logic of geopolitical conflict. Thus, the economist's traditional concern with people's access to scarce resources was replaced by a concern with the scarcity of the knowledge of the resources necessary for their survival. "The haves versus the have nots" was replaced by "the knows versus the know nots." The key cyborg moment occurred with a reinterpretation of a basic principle of capitalism's founding theorist, Adam Smith. Smith's defense of the "invisible hand" of the market rested on the distribution of economic knowledge across agents whose certainty of epistemic access is limited to their own situations. Whereas Friedrich Hayek and the Austrian school of economics left matters in Smith's original 18th-century terms, in the hands of the Keynesians funded by the Cowles Commission, such micro-level epistemic limitations were translated into macro-level uncertainty that could be managed only by a higher-order agent, the central banker, who knows more than ordinary economic agents (cf. Fuller 1993: 74–82). The history of 20th-century logic from Bertrand Russell's theory of types to Kurt Gödel's incompleteness theorem provided the philosophical backdrop to this line of thought.

Although STS researchers normally regard Donna Haraway (1990) as the "cyborg anthropologist" *par excellence*, she was neither the first nor perhaps even the most interesting person to merit the title. A very worthy pretender to the crown has just been mentioned: Gregory Bateson (1904–80), son of the English importer of Mendelian genetics, William Bateson, and husband of the popular anthropologist, Margaret Mead. Bateson also enjoys the dubious honor of having been the person who launched the reputation of the literary agent to the scientific stars, John Brockman, who managed to get the mass trade publisher, Bantam Books, to issue what turned out to be Bateson's intellectual last will and testament, *Mind and Nature: A Necessary Unity* (Bateson 1979). This solidified Bateson's reputation as a New Age guru who by that time had settled at the newly established campus of the University of California at Santa Cruz, which is coincidentally Haraway's own academic seat.

Bateson made an early reputation as what we would now call a "reflexive ethnographer." In *Naven* (1936), a monograph on a New Guinea tribe, Bateson (and Mead) created an unprecedented photographic record of native life. However, to his credit, Bateson realized that his very positioning of the camera, not to mention his written record of what he observed and understood, provided more a map of his own mind than that of the natives. Moreover, Bateson was not afraid to generalize about the cognitive processes involved, since he realized that the natives were studying him just as much as he them. His insight is encapsulated in the concept of *schismogenesis*, literally the generation of differences. Bateson's point, which guided all his subsequent work, was that learning is always a process of distinguishing oneself from the environment, which may have either benign or malign consequences, depending on the context, which itself always changes as a result of the learning experience. Indeed, an important source of such malign consequences is a failure to register the resulting shift in the frame of reference. As popularized by Cold War operations research, Bateson was thus preoccupied with the problem of *positive feedback*.

This problem can be posed most abstractly in terms of communications theory, which defines the meaning of a transmitted message in terms of the change it produces in the receiver's knowledge base: the bigger the change, the more informative the message. Now, as this process is reiterated over time, the receiver must try harder and invest more resources to find comparably informative messages – or else halt the inquiry, having decided she has learned enough to achieve equilibrium with the environment. Bateson was struck by how *rarely* human systems took the latter route of *negative feedback*, even though it was often warranted by the cost and potential risk involved in indefinitely reiterating one's earlier efforts. However, biological systems paid attention to negative feedback all too well. Darwin's scientific rival, Alfred Russel Wallace, had already compared natural selection to a steam engine's governor, which disposed of the lingering image of nature in "natural selection" operating with human-like deliberation (Bateson 1979: 195). The implications of this self-regulating image of the biosphere for development policy were bracing: cyborg equilibrium may entail a Malthusian state in which poverty and war are rationalized as compensation for overpopulation. Thus, a cost of merging the natural and the artificial into a global cyborg – nowadays, after the chemist James Lovelock, called "Gaia" – is that distinctly human behaviors are potentially disposable as system pollutants: not only may species parts be interchangeable, so too parts of species populations, including our own, when planet Earth is treated as a system ever in search of equilibrium.

Bateson had been inspired by Lewis Fry Richardson, a British meteorologist whose mathematical modeling techniques anticipated modern catastrophe and chaos theory. A devout Quaker writing in the wake of World War I, Richardson argued that the prospects for world peace were undermined by opponents whose mutual fear and distrust impeded their ability to recognize systemic changes that resulted from the actions they reinforced in each other. While it might make sense for a weaker party to aspire to military parity with its stronger opponent, once both parties are in comparable positions to harm each other, it no longer makes sense to produce more arms. Nevertheless, Richardson hypothesized that even once opponents reached such a state of "mutually assured destruction" (MAD), it was unlikely that they would stop trying to catch up. This meant that, assuming that actual warfare did not break the cycle, the arms race would be won by the party who managed to outspend the other. Of course, all of this expenditure would be a waste, exacting untold cost on the drained economies. Bateson's innovation here was to interpret this situation, which came to epitomize the systemic irrationality of the Cold War, in biological and psychological terms – a case of an *adaptation* metamorphosing through a series of step-changes into its very opposite, an *addiction*.

Richardson's concerns were taken up in the US at the outset of World War II in a series of now famous Macy Conferences, named for a Quaker shipping family that numbered among the country's founding fathers. Bateson was one of the social scientists who attended these conferences, organized at Princeton by Wiener and his mathematical colleague John von Neumann, the father of game theory (Lipset 1980: 178–83). Much of their discussion focused on the theory of "logical types," which Bertrand Russell had introduced at the dawn of the 20th century to solve the "liar's paradox," which forces one to decide whether a Cretan who says "all Cretans are liars" is telling the truth or not. Russell's solution amounted to distinguishing the speech-act itself (which asserts the Cretan's own veracity) from its referent (which points to the lying nature of Cretans). Such a distinction presupposes the ability to adopt a "second-order perspective" whereby the truth-value of one's own statements can be regarded from a context detached from the context of utterance. After Quine (1953), this has been portrayed as a "semantic ascent," which captures the sense in which a Cretan capable of observing that all Cretans are liars has effectively stepped out of her role as Cretan. Expressed in mathematical terms, what had previously been treated as a constant (one's indelible identity as a Cretan and a liar) is now rendered as a variable (someone who, in some but not necessarily all contexts, is identified as a Cretan and a liar).

Bateson became convinced that many tensions in contemporary life – from schizophrenia to interpersonal problems to global conflict – could be resolved by attending to logical types and hence breaking the "vicious circle" (a Russellian coinage) to which Richardson had drawn attention. In practice, this would involve recognizing the interdependence of the object's identity and the subject's standpoint. Bateson was fond of equating this recognition with the realization that there is never a "definitive" mapping of a territory but only different mappings, each relative to the mapper's frame of reference. A "theory" is then all the maps that are projectible from a common set of physical parameters that define the theory's syntax. In contemporary philosophy of science, this perspective is associated with the so-called "semantic" approach to theories that has received a broad endorsement among philosophers sympathetic to STS, most notably Ronald Giere (1999). It permits the prospect that what superficially appear to be two contradictory theories are, on closer inspection, alternative projections – or "models" – of the same theory, both of which may be equally adequate to the phenomena but articulated from different perspectives or designed for different purposes. This reinterpretation would then open the door to greater coordination, cooperation, or at least communication, as all sides realize that they are dealing with a common reality and, at least in that sense, implicated in a common fate.

As it turns out, Conrad Waddington was among the first to see the relevance of Bateson's cyborg anthropology to the organization of science. Already amid Bernal-inspired discussions of planning for both science and society, Waddington (1944: ch. 9) appealed to schismogenesis for how science could generate internal rivalry without detrimental spillover effects in the larger society. In effect, science's relative autonomy from society ensured that the logical difference between map and territory was maintained. Interestingly, Waddington failed to recognize that science's increasing demands on society's material resources for the conduct of its own affairs – what Derek de Solla Price would start calling "big science" in the 1960s – would place the maintenance of that boundary under severe strain. Indeed, STS is arguably the understanding of science and technology that results once the blurring of logical types is no longer seen as in need of diagnosis but simply accepted as fact. This would certainly explain the relationship between Bateson and Donna Haraway, whose vast overlap in substantive interests is overshadowed by a profound difference in attitude that I call, in deference to Hayden White (1972), *tragic* and *comic*.

Bateson's tragic vision was born of an analytic mindset founded on the realization that there are no unmitigated goods: what produces good consequences in the short term or the small scale may be bad from a long-term or large-scale perspective (and vice versa, of course). Russell's theory of types is alert to this prospect, a secular version of the Catholic doctrine of "double effect," itself a behaviorist spin on the classical Greek *hamartia* (i.e. the character trait of the tragic hero that is responsible for both his initial heroism and eventual tragedy). Bateson recognized the illusory relief provided by the MAD strategy in the Cold War arms race, whereby a policy of "deterrence" might unwittingly at any moment trigger a nuclear war.

In contrast, Haraway's comic vision reflects someone who conceptually inhabits a world where the nuclear threat never materialized. The reference to comedy is meant to recall the theory – common to Kant, Spencer and Freud – that laughter descends from the primitive release of breath from pent-up emotions with the passage of a potential danger. For Haraway, the cyborg is not a self-regulating machine in which humanity is trapped but a figurative space in which humanity can escape the constraining binaries of class, gender, race, and even species. Whereas Bateson treated the power of computers to "virtualize" reality as a threat to our ontological security, Haraway remains more sanguine about its emancipatory potential. "Modernist" *versus* "postmodernist" only begins to capture the difference in sensibility here, which will be pursued in the next and final chapter. Haraway is widely quoted as having claimed that she would rather be a cyborg than a goddess. It would seem that simply being a human is no longer an option – or so it would have seemed to Bateson.

7

The Future of Science and Technology Studies

1. What Has STS Learned from the History of Science?

Science and Technology Studies (STS) represents a distinct convergence of the history, the philosophy, and the sociology of science and technology. STS's historical dimension provides a basis for learning from science's past to enable better decisions about its future. Indeed, if the history of science teaches just one lesson, it is that today's leading scientific theories are fatally flawed – though it will probably take a while to discover how. (Compare accounts of the state of science in works published today and 100 years ago.) I deliberately use the expression "fatally flawed" to hark back to Greek tragedy, in which the hero falls by displaying a trait that had previously served him well but ends up undermining him in the course of the drama. Moreover, also like the tragic hero, the falsified scientific theory is ultimately redeemed, albeit under somewhat diminished circumstances. Thus, when Einstein's theory of relativity displaced Newtonian mechanics, the latter lost its status as the universal theoretical foundation of physical reality but retained its power to explain and predict bodies moving at speeds well below that of light. While Newtonian mechanics does not provide an adequate account of the divine plan (which had been Newton's intention), it still covers virtually all of physical reality that matters to ordinary human beings: not a bad consolation prize.

Thoughts of "falsification" naturally turn to what Francis Bacon originally called a "crucial experiment," whereby the incumbent and a challenger theory make contradictory predictions about what will happen in a given situation. This image works well if science is envisaged as a kind of highbrow game, in which prior track record may lead one to prefer the incumbent but in the end the victor is decided by the actual match. To be sure, such an image was plausible as long as resources were not so concentrated in the incumbent theory that its dominance became overpowering. Indeed, the 18th century was full of prize-induced incentives for clever people to challenge the current scientific orthodoxies. However, starting in the second half of the 19th century, as science came to be seen as a bulwark in geopolitical strategy, research and teaching came to be nationally consolidated, as symbolized by the introduction of textbooks whose authority extended well beyond their authors' universities. It gradually became difficult to imagine how an incumbent theory could be challenged other than from within, given that it had effectively received a state license to reproduce itself in successive generations of students. And the longer it has taken for scientists to establish their independent standing as researchers, the likelihood diminishes that they would bite the hand that has fed them for so long. Thus, precisely as Kuhn (1970) supposed, one increasingly waits until a dominant theory collapses from under its own conceptual and empirical deficiencies before a serious shift in allegiances occurs.

What I have just described does not simply encapsulate the last 200 years of the history of science but also explains the shift in intuitions regarding the philosophy of science – from the logical positivists and the Popperians, on the one hand, to Kuhn and his followers in the first generation of STS, on the other. As the social organization of science has changed historically, the relative plausibility of philosophical images of science has shifted accordingly. My own work in social epistemology has consistently drawn attention to this point (e.g. Fuller 1988: ch. 7).

Given the changes in political economy and intellectual focus that science has undergone with the end of the Cold War – namely, greater decentralization and biologization of the research agenda – how does and should STS position itself today? During the Cold War, it was common to say that science had finally globalized humanity. It was not that science had come to be pursued "by the people and for the people," to recall Abraham Lincoln's rhetorical definition of democracy. Rather, it was that through science, all of humanity, for the first time, was subject to a common threat. This threat had both its sources and its solutions rooted in science. Gradually the exact nature of the threat metamorphosed from mutually assured nuclear destruction to some sort of catastrophic climate change. (And there are some, like James Lovelock of Gaia hypothesis fame, who believe that the worst effects of climate change might be forestalled by a recommitment to nuclear energy.) Perhaps, in the future, science's global reach will be symbolized by an ongoing dialectic of planned (e.g. by terrorists) and unplanned (e.g. by free trade) pandemics.

This prognosis would seem grim, even cynical, were we not mindful that vague, long-term, widespread threats have been historically a spur to basic research. The real problem with science's self-globalizing momentum is that science becomes a victim of its own success. In other words, science increases our capacity for identifying and controlling for differences. This point appears in the market-driven character of the biomedical sciences, where research in potentially lucrative areas swamps inquiries that can claim only theoretical interest. The result is a customized scientific enterprise that conflates sponsorship with validation, collapsing science and technology into an undifferentiated knowledge-power regime of "technoscience," that talismanic word in the STS lexicon.

2. Has STS Succeeded at the Expense of Science and Politics?

By certain obvious standards, STS has made progress over the past quarter-century. STS is now clearly defined by a cluster of well-established journals, whose articles are increasingly recognized outside the field's borders (by one count, Bruno Latour is the fifth most highly cited social scientist overall). If STS departments have not spread as much as one might wish, their graduates have nonetheless succeeded both in and out of academia. Moreover, for policy-makers virtually everywhere, STS is seen as a uniquely desirable competence for coming to grips with technoscience in contemporary society. In retrospect, the "Science Wars" that closed the 1990s, while uncomfortable for a few of us, have managed – however unwittingly – to impress upon the general public the need to renegotiate the social contract of science and technology that went unquestioned during the Cold War. All good news for STS, it would seem.

However, in one respect STS has not really proved itself: *are we a force for good in the world?* Recall a slogan from the 1960s: "If you're not part of the solution, you're part of the problem." Today's STS researcher would no doubt object: "How asymmetrical! How beholden to dichotomous thinking!" Please forgive me while I resist such a knee-jerk response and excavate the ancient discourse that distinguishes clearly between good and bad consequences.

Consider what it has meant for Western knowledge producers to have operated in the shadow of the Cold War for the past half-century. On the one hand, in the name of consolidating both national and international knowledge bases, the Cold War seeded processes for the exchange, evaluation, and integration of knowledge that carry on as the infrastructure for post-Cold War knowledge policy. One need only recall the cover of "national security" under which the Internet, the *Science Citation Index*, and artificial intelligence research were first developed. On the other hand, the withdrawal of what Alvin Gouldner (1970) dubbed the "welfare-warfare state" from regulating these developments has left them captive to the competing demands of the global marketplace and sectarian communities, as exemplified by the uncertainties surrounding the institutional future of the university in most countries. It might be said that the day-to-day autonomy routinely granted by the state to researchers during the Cold War exacted a high cost: there was no incentive – indeed, when national security was at stake, there was outright prohibition – for researchers to address for themselves the ends of their knowledge production. Consequently, today there is little, if any, rhetorical space available for expressing a research standpoint that is autonomous from the interests of potential clients that does not also appear like a self-serving plea for the researcher's narrow guild interests.

During the Cold War, all self-proclaimed progressive thinkers were touched by the hand of Marx. There was a clear sense of where all of "us" were supposed to be heading. Of course, Liberals, Social Democrats, and Communists differed over the means, the speed, and the distribution of the burden to complete the task at hand. But the ultimate goal was largely the same: a society where everyone could fulfill their potential without inhibiting the ability of others to do the same – and importantly, where everyone recognized the desirability of this goal. Jerome Ravetz's (1971) classic *Scientific Knowledge and Its Social Problems* epitomized this pre-history to STS. STS as we now know it began to acquire an institutional presence as this consensus started to be questioned – and really gained momentum once the consensus was blown apart in the 1980s.

Socialism's setbacks – ranging from the fiscal crisis of the welfare state to the collapse of the Soviet Empire – led to a profound questioning of not only political allegiances but also politics altogether. Along with skepticism about the meaningfulness of "left" versus "right" came a withdrawal of the younger generation – that is, those who have come of age since around 1980 – from the party politics and electoral processes that had underwritten the left–right polarity for two centuries. Perhaps the subtlest manifestation of this shift has been the semantic dissipation of political language: nowadays just about everything is "political" *except* the practices conventionally associated with politics. Thus, the incorporation of non-human agents into academic and popular narratives is "political," as are the attempts by various well-organized interest groups to have their lifestyles recognized and secured. The hand of STS in these developments is, of course, unmistakable.

What STS has helped politics to lose – a sense of *res publica* or the "common weal" – is matched by its complicity with the "knowledge society" discourse that purports to say that knowledge is produced just about everywhere *except* universities, which are now reduced to "trading zones" (Nowotny et al. 2000). Notwithstanding Latour's (1993) obfuscations about the "a-modern" and the "non-modern," the program he assembled at the 2004 meeting of the Society for Social Studies of Science in Paris clearly fulfilled Lyotard's (1983) postmodern prophecy of an increasingly dispersed knowledge society. Gone is the idea of knowledge that may be made universally available from a central academic location. Moreover, there are no reflexive surprises here. STS practitioners do not merely represent, perform, or "give voice" to groups traditionally excluded by democratic politics. They themselves are often academically marginal – the contract-based teachers and researchers that Daryl Chubin once dubbed the "unfaculty." But even the more successful members of the field continue to harbor resentment of traditional academic structures, alongside cynicism to "politics as usual." As idea or institution, politics and academia are portrayed as parasitic and peremptory over the full range of what is increasingly given such generic but no less metaphysically freighted names as "agency" or "life." It would not surprise me if before long the word "knowledge" is dropped from the STS vocabulary for ranging too prescriptively over the available resources to get action.

3. Anticipating Future Historical Judgments of STS

My guess is that future historians will find four things behind this perspectival shift, which I present in order of increasing generality:

1. The resentment of the younger generation toward older politicians and professors who have abused the offices to which the young

themselves may have once aspired. This is most clearly manifested in skepticism toward the institution of lifelong academic tenure, which was designed to encourage the pursuit of inquiry with impunity but has too often instead led to lack of public account-ability and intellectual stagnation.

2. An intolerance for the level of error that invariably arises from pur-portedly "free" action and inquiry. After all, permitting politicians and academics a free hand in their dealings has just as often led to outright disasters as mere stagnation or squandered opportunities. Thus, the pendulum swings back, resulting in a collapse of the temporal distinction between a policy proposal and its subsequent correction (or reversal) into an undifferentiated state of "ambiva-lence," whereby if you don't already know in advance that good will be produced, you are hesitant to try at all. This line of think-ing lay behind the "precautionary principle" and, more generally, the "risk society" – both of which have received strong support from STS.

3. Such impatience with learning from mistakes reflects an intellec-tual captivity to the accelerated pace of life – or "addiction to speed," as Paul Virilio would say. It tends to conflate diminishing returns on investment with the outright conversion of benefits to harms. For example, the fact that women have not made as much social progress in the last decade as they did in the next-to-last decade is taken to mean that the original strategy had been based on a faulty "totalizing" conception of gender, when it may simply point to the need for greater tactical sophistication at a practical level to achieve the ultimate goal. One can agree with Wittgenstein that the meaning of a word like "gender" does not dictate its full usage without concluding that the word is useless. Yet, the ease with which STS relinquishes "universalist" ambitions in this fashion is symptomatic of a loss of transgenerational memory that forgets how much worse things were not so long ago. Such collec-tive amnesia is only abetted by STS's studied anti-institutionalism.

4. Last but not least is the ongoing psychic need to make the most of the hand that fate has dealt, which leads each new generation to find silver linings in even the darkest clouds. Jon Elster (1983) coined the useful phrase "sweet lemons" (the converse of "sour grapes") for this phenomenon. Once one loses faith in the tradi-tional academic and political structures, the Wittgensteinian aim of "leaving the world alone" starts to look less bad than any more activist alternative. In this context, "radicalism" amounts to little more than allowing what had been previously hidden to reveal

itself. This manner of speaking, reminiscent of Heidegger's conception of truth as *aletheia*, effectively divests the analyst of any responsibility for what is revealed, since, as Latour says, the STS practitioner is merely "following the actants."

To be sure, this passive yet receptive stance enables STS to exert a tactical advantage over rival empirical analysts, notably Marxists, whose conceptual framework is so normatively loaded that it cannot register certain facts without passing judgment on them. From an STS standpoint, this results in wildly unreliable observations that tend to over- or undervalue phenomena, depending on whether they are deemed "emancipatory" or "exploitative," "progressive" or "reactionary" – or even whether the object of inquiry is identified as a "producer" or a "consumer." Whereas the client of Marxist analysis is likely to come away feeling guilty for her complicity in the situation that led to the employment of the Marxist's services in the first place, the STS client simply feels epistemically enhanced and welcomes another opportunity to learn more in the future. However, the history of client-driven ethnography has shown that the client's enlightenment typically comes at the expense of the power that the object of inquiry had exerted through its obscurity. Depending on the original balance of power between client and object, this may or may not be a desirable outcome. STS practitioners do not care to decide the matter because they never know who might require their services in the future.

A consequence of STS's professionalized value neutrality is a profoundly distorted understanding of the equation "knowledge is power." In its original Enlightenment form, the equation was meant to express that the more we know, the less power others have over us. Indirectly, of course, this implies that our own power grows – but only in the strict sense of the original Latin *potentia*, that is, the sphere of possible action grows. In this somewhat metaphysical sense more knowledge makes us "more free," but that entails new burdens as suddenly we are faced with a larger decision space than we had previously envisaged. It is worth recalling that the opposing model was one of religious leaders who ruled by promulgating dogmas that discouraged people from trying to move in new directions. These leaders were able to get their way not because they could enforce their will in all cases but because, in most cases, the rest of society already believed that nothing could be done to oppose them and hence nothing was tried.

I fear that STS has lost sight of this context. Specifically, "knowledge is power" was meant to *loosen*, not tighten, the relationship between knowledge and power, especially in its ordinary sense of "the capacity to get action." In contrast, much of the attraction of STS research to both academics and policymakers comes from revealing intermediaries whose distinctive knowledge enables them to act in ways that have perverse effects on the schemes promoted by nominally more powerful agents. The implicit explanation for this phenomenon is that the intermediaries live radically different lives from the would-be hegemons. From this, STS infers that power resides in one's specific epistemic embodiment. However, from an Enlightenment standpoint, a social epistemology based on such an "indigenized" conception of knowledge is simply a recipe for democratizing rule by high priests – that is, many more priests ruling over much smaller domains.

STS's all too open normative horizons of inquiry would have been criticized in the past for its "mere instrumentality," "rank opportunism," and "lack of clear theoretical focus." However, nowadays it is given a philosophical high gloss with the help of some funky process metaphysics inspired by two outliers in 20th-century philosophy, Henri Bergson and Alfred North Whitehead (Latour 2004). A sociologist of STS knowledge – a social epistemologist – would relate this shift in status to the lack of institutional protection for the vast majority of STS practitioners today. That academics could enforce their will on the rest of society may have been always an illusion, but at least they used to exert sufficient control over their own sphere of activity to launch a consistent critique. But what happens once the prospect of a stable counterpoint vanishes? Process metaphysics, whatever its other merits, is the ideological expression of researchers in perpetual need of shoring up their optimism as they remain in profound ignorance about the source of their next paycheck. In that context, whatever appears on the horizon is easily interpreted as a harbinger of good things – at least until the next phase in the process emerges.

That STS practitioners nowadays root around in the metaphysical marshlands of Bergson and Whitehead to persuade themselves of these semantic maneuvers suggests a deep state of denial about how our field might be judged. After all, the quickest theoretical route to most of the "radical" things STS wants to say these days about the "distributed" and "emergent" nature of technoscience is through the work of Friedrich Hayek, who provided a systematic metaphysical underpinning to the market that would have made even Adam Smith blush. Of course, Hayek wore his politics – or I should say his antipolitics (he was one of the modern skeptics about the value of elections) – very much on his sleeve. This explains Hayek's influence on the likes of Reagan, Thatcher, and Pinochet. It equally explains why our ever politically correct field has been reluctant to embrace Hayek. After all, we self-avowed "radical" thinkers in STS would hate to think that *we have always already been neo-liberal*. But future historians may judge otherwise – and more harshly. We may turn out to have been part of "the problem" rather than "the solution" that institutions like the university and movements like socialism – however inadequately – tried to address. In any case, future historians will find STS's tragically hip turns of thought an endless source of insight about the overall mindset of our times.

So, if the question is not too grandiose, what has been STS's contribution to world civilization? At the start of my career, in the early 1980s, I would have said that STS contributes to the second moment of a dialectic that aims to realize the Enlightenment dream of truly universal knowledge. I read works like Bloor (1976) and Latour and Woolgar (1979) as revealing the captivity of normative philosophy of science to wishful thinking about the history and sociology of science. Philosophers wrote as if scientists were trying to live up to their normative ideals, even though the philosophers themselves could not agree on what those ideals were. STS showed that philosophers suffered less from bad faith than sheer credulousness. They – and such sociological fellow-travellers as Robert Merton – made the fatal mistake of believing their own hype. Like overzealous imperialists, philosophers failed to recognize the "made for export" quality of their own normative discourse. Put crudely, the "scientific method" had more impact in disciplining school children and regimenting the nonsciences than in regulating the practices of real scientists.

My version of "social epistemology" has been dedicated to bridging this very significant gap between the "is" and the "ought" (Fuller 1988). It has increasingly led me to consider the conditions under which knowledge is institutionalized as a public good – that is, an entity capable of benefiting the vast majority not involved in its production. In the current STS Newspeak, this idea is unthinkable since all knowledge is necessarily "co-produced." Thus, the sorts of problems economists have traditionally associated with public goods, such as the "tragedy of the commons," can be dismissed as simply efforts to demean the ingenuity of our successors to interpret the potential of the commons rather differently from us. Undeterred by such sophisms, I have been led to champion both the classical university and more new-fangled consensus conferences, while opposing the excesses of "evidence-based policy" and "knowledge management," to name two fields that capitalize on the rhetoric of "access," "inclusion," and even "democratization" only to deskill inquirers and debase organized inquiry (Fuller 2000a, 2002a, 2006a: ch. 6). That STS practitioners have been all too eager to contribute to these fields may be indicative of just how much the field has fallen victim to its own success. We have had much to say about deconstruction but precious little by way of *reconstruction*.

4. Conclusion: Is STS a Symptom in the Passage of Humanity?

If nothing else, STS has proven "useful" to a wide range of constituencies. These include science policymakers trying to foster basic research, social engineers interested in maximizing the diffusion of a new technology, and, of course, line managers in search of creative accounting techniques to cut costs and boost profits. That STS has afforded so many applications, despite its equally many theoretical controversies, demonstrates just how eminently detachable our field's practices are from its theories. A Marxist devoid of charity might even suggest that the more exotic theories by Latour and Haraway, to which the entire field of STS tends to be reduced by its many admirers in cultural studies, are little more than "ideological superstructure" on the soft-focus surveillance that characterizes so much of STS empirical work. Having now evaluated scores of grant proposals, academic manuscripts, doctoral dissertations, and candidates for tenure and promotion in several countries, I must confess some sympathy with our Marxist's jaundiced judgment. There is probably no other field whose members are so adaptable to circumstance. STS is a veritable cockroach in today's intellectual ecology – the ultimate compliment, some evolutionists might say.

STS offers something for everyone – something funky for the high theorists blissfully ignorant of the workings of technoscience and something more nuts-and-bolts for harried decision-makers pressed to justify hard choices. What STS lacks is a unity of purpose, a clear sense of how its empirical work is, indeed, an application – or, still better, a test – of its theories. My concern with this absence of unity may strike you as a regrettable modernist hang-up. Nevertheless, it may also help to explain why academic institutionalization has continued to elude STS. Moreover, in this respect, STS would not be unique. Devotees of the popular science literature read much of "chaos" and "complexity," which refer to certain mathematical properties of natural and artificial systems that have become easier to detect and represent in recent years because of technical advances in computer simulation. Many quite unrelated phenomena can be modeled now as chaotic and complex systems. But do they add up to a coherent world-view? Contrary to the extravagances of science popularization, they don't. Chaos and complexity remain no more than a sub-set of the forms of quantitative analysis available for the conduct of normal science in a variety of disciplines. Something similar may be said about STS vis-à-vis the larger knowledge system.

Just like chaos and complexity, the leading so-called "theory" in STS, actor-network theory, is not really a theory at all but an all-purpose *method* (Law 2004). Daniel Dennett (1995), aping William James, has dubbed Darwin's theory of natural selection the "universal solvent." Actor-network theory works its methodological magic in just this way: you think you can isolate who or what is responsible for the larger effects in our technoscientific world? Well, that must mean you haven't accounted for all the "actors." Once you've done this, you'll realize that "agency" is distributed across a wide range of entities that transgress the usual ontological categories, not least human/non-human. After all, agency is no more than the capacity to act. When you take this point seriously, you'll resist the urge to jump to conclusions about the ascription of either "blame" or "credit." Such moral qualities need to be spread more evenly, and hence more thinly, across a wider range of entities. Make no mistake: this is *not* anthropomorphism. It is almost the exact opposite. Actor-network theorists aren't trying to attribute properties to, say, scallops and doorstops that were previously restricted to humans. Rather, they want to convert what these entities normally do into the benchmark of agency. The unspoken implication is that what remains distinctive to humans is valued less. Indeed, the unique displays of intelligence that have enabled us to dominate nature may be profligate – the metaphysical equivalent of industrial pollution in some fanciful "political ecology," the field which Latour (2004) nowadays tries to represent.

To be sure, STS did not originate this line of thought. But because the field travels with such light theoretical baggage, it is better placed than most to capitalize on it. Observed from this moment in intellectual history, STS is turning out to be the crucible in which the two great antihumanist trends of the late 20th century are being brought together in a heady stew with potentially explosive scientific and political consequences. The first is, broadly speaking, post-structuralist thought, rooted in Nietzsche and Heidegger and brought to fruition in 1960s France in works by Roland Barthes and Michel Foucault that announced the death of something variously called "the author," "the subject," or simply "man." The second stream of anti-humanism stems from the technologically enhanced version of neo-Darwinism that has come to dominate science policy and the public understanding of science in the post-Cold War era. Together they call into question the uniqueness of *humanity*, not least by displacing, if not actually disparaging, the two modern projects that have stood most clearly for that concept: *social science* and *socialism* (Fuller 2006b). Are we sure we want STS to be remembered as such a strong supporting player in this trajectory?

A sign of the times is that the literary agent John Brockman, one of the most important figures in the intellectual world today, has appropriated "the third culture" – an old phrase for the social sciences – to cover an interdisciplinary broad church devoted to recovering the "nature" in "human nature." Richard Dawkins, E. O. Wilson, and Steven Pinker are just some of the names associated with Brockman's project, conveniently located at the website, <www.edge.org>, not to mention the many popular books he has helped to publish over the past quarter-century, starting with the original cyborg anthropologist, Gregory Bateson (1979). STS's own "made for export" gurus, Bruno Latour and Donna Haraway, are unacknowledged affiliates. Latour has never hidden his belief that a preference for "biosociality" over "sociobiology" is ultimately a matter of political correctness among English speakers. He is happy to live with either or both. As for Haraway, she spent the better part of the 1990s deconstructing postwar attempts to deconstruct UNESCO's attempt to forge a biological unity for humanity (esp. Haraway 1997: ch. 6). In recent years, she has beaten a hasty cynic's retreat, literally going to the dogs in her latest tract, *A Companion Species Manifesto* (Haraway 2003). Wading through her endless name-checks and abuses of the subjunctive mood, one gets the impression that Haraway really does believe that the best way to understand the human condition is by studying our relations with canines. This might work for humans who have already exhausted the more obvious modes of inquiry, or for Martians in search of the exact distinction between humans and dogs. But we don't live in either world. It is one thing for STS to be inspired by science fiction, quite another to become science fiction.

We are entering an era that may be remembered for its *casualization of the human condition* (Fuller 2006b: 12). Technological advances are making it easier for people to come in and out of existence, to which STS makes its own conceptual contribution by manufacturing a discourse that facilitates the exchange between human and non-human qualities. Add to this the emergence of Peter Singer as the world's leading public philosopher, who now calls for a politics of the left that replaces Marx with Darwin as its standard-bearer. Singer (1999) sees it in terms of an expansion of the moral circle to cover all forms of life. His utopian vision presupposes that we have already closed the moral circle around all forms of human life. Yet the growing disparity between the rich and the poor – both between and within countries – testifies otherwise. Many in STS are attracted by utopian politics, pretending that we are much farther along in history than we really are. At least, Singer displays the courage of his convictions by providing arguments for why humans should make room for non-humans by limiting and even forgoing their own lives. STS certainly knows how to talk the talk. But does it dare walk the walk?

Bibliography

Agassi, J. (1983). "Theoretical Bias in Evidence: A Historical Sketch," *Philosophica* 31: 7–24.

Agassi, J. (1985). *Technology: Philosophical and Social Aspects*. Dordrecht: Kluwer.

Ashmore, M. (1989). *The Reflexive Thesis*. Chicago: University of Chicago Press.

Babich, B. (1997). "The Hermeneutics of a Hoax: On the Mismatch of Physics and Cultural Criticism," *Common Knowledge* 6/2 (September): 23–33.

Baker, K. (1975). *Condorcet: From Natural Philosophy to Social Mathematics*. Chicago: University of Chicago Press.

Barron, C. (ed.) (2003). "A Strong Distinction between Humans and Non-Humans is No Longer Required for Research Purposes: A Debate between Bruno Latour and Steve Fuller," *History of the Human Sciences* 16/2: 77–99.

Basalla, G. (1967). "The Spread of Western Science," *Science* 156 (5 May): 611–22.

Basalla, G. (1988). *The Evolution of Technology*. Cambridge, UK: Cambridge University Press.

Bateson, G. (1979). *Mind and Nature: A Necessary Unity*. New York: Bantam.

Baudrillard, J. (1983). *Simulations*. New York: Semiotexte.

Bechtel, W. (ed.) (1986). *Integrating Scientific Disciplines*. Dordrecht: Martinus-Nijhoff.

Beniger, J. (1986). *The Control Revolution: Technological and Economic Origins of the Information Society*. Cambridge, MA: Harvard University Press.

Benjamin, W. (1968). "The Work of Art in the Age of Mechanical Reproductions in H. Arendt (ed.), *Illuminations* (orig. 1936). New York: Harcourt Brace & World, pp. 217–52.

Bergson, H. (1935). *The Two Sources of Morality and Religion* (orig. 1932). London: Collier Macmillan.

Berlin, I. (1969). *Four Essays on Liberty*. Oxford: Oxford University Press.

Biagioli, M. (ed.) (1999). *The Science Studies Reader*. London: Routledge.
Blomstrøm, M., and Hettne, B. (1984). *Development Theory in Transition*. London: Zed.
Bloor, D. (1976). *Knowledge and Social Imagery*. London: Routledge.
Brannigan, A. (1981). *The Social Basis of Scientific Discoveries*. Cambridge, UK: Cambridge University Press.
Bush, V. (1945). *Science: The Endless Frontier*. Washington, DC: US Government Printing Office.
Byrne, R. and Whiten, A. (eds.) (1987). *Machiavellian Intelligence: Social Expertise and the Evolution of Intelligence in Monkeys, Apes and Humans*. Oxford: Oxford University Press.
Callebaut, W. (ed.) (1993). *Taking the Naturalistic Turn*. Chicago: University of Chicago Press.
Callon, M. (1986). "Some Elements of a Sociology of Translation," in J. Law (ed.), *Power, Action, and Belief*. London: Routledge & Kegan Paul, pp. 196–229.
Cardwell, D. S. L. (1972). *Technology, Science and History*. London: Heinemann.
Carley, K. and Kaufer, D. (1993). *Communication at a Distance*. Hillsdale, NJ: Lawrence Erlbaum Associates.
Cassirer, E. (1923). *Substance and Function* (orig. 1910). La Salle, IL: Open Court Press.
Cassirer, E. (1950). *The Problem of Knowledge: Philosophy, Science, and History since Hegel*. New Haven: Yale University Press.
Castells, M. (1996–8). *The Information Age: Economy, Society and Culture*, 3 vols. Oxford: Blackwell.
Cat, J., Cartwright, N. and Chang, H. (1996). "Otto Neurath: Politics and the Unity of Science," in Galison and Stump (eds.) (1996), pp. 347–69.
Ceccarelli, L. (2001). *Shaping Science with Rhetoric: The Cases of Dobzhansky, Schrödinger, and Wilson*. Chicago: University of Chicago Press.
Cohen, H. F. (1994). *The Scientific Revolution: An Historiographical Inquiry*. Chicago: University of Chicago Press.
Cohen, I. B. (1985). *Revolution in Science*. Cambridge, MA: Harvard University Press.
Cohen, I. B. (1995). *Science and the Founding Fathers*. Cambridge, MA: Harvard University Press.
Collins, H. and Yearley, S. (1992). "Epistemological Chicken," in A. Pickering (ed.), *Science as Practice and Culture*. Chicago: University of Chicago Press, pp. 301–27.
Collins, R. (1998). *The Sociology of Philosophies: A Global Theory of Intellectual Change*. Cambridge, MA: Harvard University Press.
Commager, H. S. (1978). *The Empire of Reason: How Europe Imagined and America Realized the Enlightenment*. London: Weidenfeld and Nicolson.
Conant, J. B. (1950). *The Overthrow of the Phlogiston Theory: The Chemical Revolution of 1775–1789*, Harvard Case Histories in Experimental Science, Case 2. Cambridge, MA: Harvard University Press.

Cosmides, L. and Tooby J. (1994). "Beyond Intuition and Instinct Blindness: Towards an Evolutionarily Rigorous Cognitive Science," *Cognition* 50: 41–77.

Creath, R. (1996). "The Unity of Science: Carnap, Neurath, and Beyond," in Galison and Stump (eds.) (1996), pp. 158–69.

Culler, J. (1975). *Structuralist Poetics*. Ithaca, NY: Cornell University Press.

Culler, J. (1982). *On Deconstruction*. Ithaca, NY: Cornell University Press.

Dahrendorf, R. (1995). *LSE: A History of the London School of Economics and Political Science 1895–1995*. Oxford: Oxford University Press.

Davies, K. (2001). *Cracking the Code*. New York: Free Press.

Dawkins, R. (1976). *The Selfish Gene*. Oxford: Oxford University Press.

Dawkins, R. (1982). *The Extended Phenotype*. Oxford: Oxford University Press.

Deleuze, G. (1984). *Difference and Repetition* (orig. 1968). New York: Columbia University Press.

Dennett, D. (1995). *Darwin's Dangerous Idea: Evolution and the Meanings of Life*. New York: Simon & Schuster.

Dennett, D. (2003). "In Darwin's Wake, Where Am I?," in. J. Hodge and G. Radick (eds.), *The Cambridge Companion to Darwin*. Cambridge, UK: Cambridge University Press, pp. 357–76.

Dickens, P. (2000). *Social Darwinism*. Milton Keynes: Open University Press.

Dobzhansky, T. (1937). *Genetics and the Origin of Species*. New York: Columbia University Press.

Dorn, H. (1991). *The Geography of Science*. Baltimore: Johns Hopkins University Press.

Duff, A. (2005). "Social Engineering in the Information Age," *The Information Society* 21: 67–71.

Dupré, J. (1993). *The Disorder of Things*. Cambridge, MA: Harvard University Press.

Dupré, J. (2003). *Darwin's Legacy: What Evolution Means Today*. Oxford: Oxford University Press.

Durkheim, E. (1952), *Suicide* (orig. 1897). New York; Free Press.

Durkheim, E. (1964), *Rules of the Sociological Method* (orig. 1895). New York: Free Press.

Durkheim, E. (1997). *The Division of Labor in Society* (orig. 1893). New York: Free Press.

Eco, U. (1976). *A Theory of Semiotics*. Bloomington: Indiana University Press.

Economist (2002). "Defending Science," 31 January.

Elster, J. (1983). *Sour Grapes*. Cambridge, UK: Cambridge University Press.

Elster, J. (1993). "Constitutional Bootstrapping in Philadelphia and Paris," *Cardozo Law Review* 14: 549–75.

Feyerabend, P. (1975). *Against Method*. London: Verso.

Feyerabend, P. (1979). *Science in a Free Society*. London: Verso.

Fleischacker, S. (2004). *A Short History of Distributive Justice*. Cambridge, MA: Harvard University Press.

Foucault, M. (1970). *The Order of Things* (orig. 1966). New York: Vintage.
Frank, A. G. (1998). *Re-Orient: The Global Economy in the Asian Age.* Berkeley, CA: University of California Press.
Frank, P. (1949). *Modern Science and Its Philosophy.* Cambridge, MA: Harvard University Press.
Franklin, J. (2000). *The Science of Conjecture.* Baltimore: Johns Hopkins University Press.
Friedman, M. (2000). *A Parting of the Ways: Carnap, Cassirer, and Heidegger.* Chicago: Open Court Press.
Fukuyama, F. (2000). *Our Posthuman Future.* New York: Straus, Farrar & Giroux.
Fuller, S. (1985). "Bounded Rationality in Law and Science," PhD. University of Pittsburgh.
Fuller, S. (1988). *Social Epistemology.* Bloomington: Indiana University Press.
Fuller, S. (1992). "Epistemology Radically Naturalized: Recovering the Normative, the Experimental, and the Social," in R. Giere (ed.), *Cognitive Models of Science.* Minneapolis: University of Minnesota Press, pp. 427–59.
Fuller, S. (1993). *Philosophy of Science and Its Discontents,* 2nd edn (orig. 1989). New York: Guilford Press.
Fuller, S. (1994). "Making Agency Count: A Brief Foray into the Foundations of Social Theory," *American Behavioral Scientist* 37: 741–53.
Fuller, S. (1996a). "Recent Work in Social Epistemology," *American Philosophical Quarterly* 33: 149–66
Fuller, S. (1996b). "Talking Metaphysical Turkey about Epistemological Chicken, and the Poop on Pidgins," in Galison and Stump (1996), pp. 170–86, 468–71.
Fuller, S. (1997). *Science.* Milton Keynes: Open University Press.
Fuller, S. (1999). "Making the University Fit for Critical Intellectuals: Recovering from the Ravages of the Postmodern Condition," *British Educational Research Journal* 25: 583–95.
Fuller, S. (2000a). *The Governance of Science: Ideology and the Future of the Open Society.* Milton Keynes: Open University Press.
Fuller, S. (2000b). *Thomas Kuhn: A Philosophical History for Our Times* (Chicago: University of Chicago Press).
Fuller, S. (2002a). *Knowledge Management Foundations.* Boston: Butterworth-Heinemann.
Fuller, S. (2002b). "Prolegomena to a Sociology of Philosophy in the 20th Century English-Speaking World," *Philosophy of the Social Sciences* 32: 151–77.
Fuller, S. (2003a). *Kuhn vs. Popper: The Struggle for the Soul of Science.* Cambridge, UK: Icon.
Fuller, S. (2003b). "In Search of Vehicles for Knowledge Governance: On the Need for Institutions that Creatively Destroy Social Capital," in N. Stehr (ed.), *The Governance of Knowledge.* New Brunswick, NJ: Transaction Books, pp. 41–76.

Fuller, S. (2006a). *The Philosophy of Science and Technology Studies*. New York: Routledge.

Fuller, S. (2006b). *The New Sociological Imagination*. London: Sage.

Fuller, S. (2006c). "France's Last Sociologist," *Economy and Society* 35: 314–23.

Fuller, S. (2007). *Science vs Religion? Intelligent Design and the Problem of Evolution*. Cambridge, UK: Polity.

Fuller, S. and Collier, J. (2004). *Philosophy, Rhetoric and the End of Knowledge: A New Beginning for Science and Technology Studies*, 2nd edn (orig. 1993, authored by Fuller). Mahwah, NJ: Lawrence Erlbaum Associates.

Galison, P. (1996). "Introduction: The Context of Disunity," in Galison and Stump (eds.) (1996), pp. 1–36.

Galison, P. (1999). "Trading Zone: Coordinating Action and Belief" (orig. 1987), in Biagioli (1999), pp. 137–59.

Galison, P. and Stump, D. (eds.) (1996). *The Disunity of Science: Boundaries, Contexts and Power*. Palo Alto: Stanford University Press.

Georgescu-Roegen, N. (1971). *The Entropy Law and the Economic Process*. Cambridge, MA: Harvard University Press.

Gerschenkron, A. (1962). *Economic Backwardness in Historical Perspective*. Cambridge, MA: Harvard University Press.

Gibbons, M., Limoges, C., Nowotny, H., Schwartzman, S., Scott, P., and Trow, M. (1994). *The New Production of Knowledge*. London: Sage.

Giddens, A. (1976). *New Rules of the Sociological Method*. London: Hutchinson.

Giere, R. (1999). *Science without Laws*. Chicago: University of Chicago Press.

Gilbert, W. (1991). "Towards a Paradigm Shift in Biology," *Nature*, (349) 10 January: 99.

Goldman, A. (1999). *Knowledge in a Social World*. Oxford: Oxford University Press.

Gorz, A. (1989). *A Critique of Economic Reason*. London: Verso.

Gouldner, A. (1970). *The Coming Crisis in Western Sociology*. New York: Basic Books.

Gouldner, A. (1976). *The Dialectic of Ideology and Technology*. New York: Seabury Press.

Grafton, A. (1990). *Forgers and Critics*. Princeton: Princeton University Press.

Grafton, A. (1997). *The Footnote: A Curious History*. Cambridge, MA: Harvard University Press.

Habermas, J. (1971). *Knowledge and Human Interests*. Boston: Beacon Press.

Habermas, J. (1981). *The Theory of Communicative Action*. Boston: Beacon Press.

Hacking, I. (1975). *The Emergence of Probability*. Cambridge, UK: Cambridge University Press.

Hacking, I. (1983). *Representing and Intervening*. Cambridge, UK: Cambridge University Press.

Hacking, I. (1984). "Five Parables," in R. Rorty et al. (eds.), *Philosophy in History*. Cambridge, UK: Cambridge University Press.

Hacking, I. (1996). "The Disunities of the Sciences," in Galison and Stump (1996), pp. 37–74.

Hacking, I. (2002). *Historical Ontology*. Cambridge, MA: Harvard University Press.

Hacohen, M. (2000). *Karl Popper: The Formative Years 1902–1945*. Cambridge, UK: Cambridge University Press.

Haraway, D. (1990). *Simians, Cyborgs and Women*. London: Free Association Books.

Haraway, D. (1997). *Modest_Witness@Second_Millennium.FemaleMan© Meets OncoMouse™*. London: Routledge.

Haraway, D. (2003). *The Companion Species Manifesto*. Chicago: Prickly Paradigm Press.

Harris, M. (1968). *The Rise of Anthropological Theory*. New York: Crowell.

Hayek, F. (1952). *The Counter-Revolution in Science*. Chicago: University of Chicago Press.

Herf, J. (1984). *Reactionary Modernism*. Cambridge, UK: Cambridge University Press.

Hirsch, F. (1976). *The Social Limits to Growth*. London: Routledge & Kegan Paul.

Hirschmann, A. O. (1977). *The Passions and the Interests*. Princeton: Princeton University Press.

Hoffman, A. (1997). *From Heresy to Dogma: An Institutional History of Corporate Environmentalism*. San Francisco: Lexington Books.

Hofstadter, R. (1965). *The Paranoid Style in American Politics*. New York: Alfred Knopf.

Holton, G. (1993). *Science and Anti-Science*. Cambridge, MA: Harvard University Press.

Horgan, J. (1996). *The End of Science*. Reading, MA: Addison-Wesley.

Inkster, I. (1991). *Science and Technology in History: An Approach to Industrial Development*. London: Macmillan.

Jarvie, I. (2001). *The Republic of Science: The Emergence of Popper's Social View of Science 1935–1945*. Amsterdam: Rodopi.

Jarvie, I. (2003). "Fuller on Science," *Philosophy of the Social Sciences* 33: 261–85.

Jasanoff, S. (2005). *Designs on Nature: Science and Democracy in Europe and the United States*. Princeton: Princeton University Press.

Johnson, P. (1991). *Darwin on Trial*. Chicago: Regnery Press.

Kitcher, P. (2001). *Science, Truth, and Democracy*. Oxford: Oxford University Press.

Klein, J. T. (1990). *Interdisciplinarity*. Detroit: Wayne State University Press.

Knorr-Cetina, K. (1999). *Epistemic Cultures*. Cambridge, MA: Harvard University Press.

Koertge, N. (ed.) (2005). *Scientific Values and Civic Virtues*. Oxford: Oxford University Press.

Kripke, S. (1980). *Naming and Necessity* (orig. 1972). Cambridge, MA: Harvard University Press.

Kuhn, T. S. (1970). *The Structure of Scientific Revolutions*, 2nd edn (orig. 1962). Chicago: University of Chicago Press.

La Follette, M. (ed.) (1983). *Creationism, Science and the Law*. Cambridge, MA: MIT Press.

Lakatos, I. (1981). "History of Science and Its Rational Reconstructions," in I. Hacking, *Scientific Revolutions*. Oxford: Oxford University Press (1981), pp. 107–27.

Lakatos, I. and Feyerabend, P. (1999). *For and Against Method*. Chicago: University of Chicago Press.

Latour, B. (1987). *Science in Action*. Milton Keynes, UK: Open University Press.

Latour, B. (1988a). *The Pasteurization of France*. Cambridge, MA: Harvard University Press.

Latour, B. (1988b). "The Politics of Explanation," in Woolgar (1988), pp. 155–76.

Latour, B. (1992). "Where are the Missing Masses? The Sociology of a Few Mundane Artefacts," in W. Bijker and J. Law (eds.), *Shaping Technology/Building Society*. Cambridge, MA: MIT Press, pp. 225–58.

Latour, B. (1993). *We Have Never Been Modern*. Cambridge, MA: Harvard University Press.

Latour, B. (1996). *Aramis, or For the Love of Technology*. Cambridge, MA: Harvard University Press.

Latour, B. (1999). *Pandora's Hope*. Cambridge, MA: Harvard University Press.

Latour, B. (2004). *The Politics of Nature: How to Bring the Sciences into Democracy*. Cambridge, MA: Harvard University Press.

Latour, B. and Woolgar, S. (1979). *Laboratory Life: The Social Construction of Scientific Facts*. London: Sage.

Laudan, L. (1996). "The Demise of the Demarcation Problem" (orig. 1983), reprinted in L. Laudan, *Beyond Positivism and Relativism*. Boulder: Westview Press, pp. 210–22.

Law, J. (2004). *After Method: Mess in Social Science Research*. London: Routledge.

Lepenies, W. (1988). *Between Literature and Science: The Rise of Sociology*. Cambridge, UK: Cambridge University Press.

Levine, D. N. (1995). *Visions of the Sociological Tradition*. Chicago: University of Chicago Press.

Lipset, D. (1980). *Gregory Bateson: The Legacy of a Scientist*. Englewood Cliffs, NJ: Prentice-Hall.

Lomborg, B. (2001). *The Sceptical Environmentalist*. Cambridge, UK: Cambridge University Press.

Lomborg, B. (ed.) (2004). *Global Crises, Global Solutions*. Cambridge, UK: Cambridge University Press.

Longino, H. (1990). *Science as Social Knowledge*. Princeton: Princeton University Press.

Lyotard, J.-F. (1983). *The Postmodern Condition* (orig. 1979). Minneapolis: University of Minnesota Press.

Machlup, F. (1962–84). *Knowledge: Its Creation, Distribution and Economic Significance*, 3 vols. Princeton: Princeton University Press.

MacIntyre, A. (1984). *After Virtue*, 2nd edn (orig. 1981). South Bend: University of Notre Dame Press.

McLuhan, M. (1964). *Understanding Media: The Extensions of Man*. New York: McGraw-Hill.

Mandelbaum, M. (1971). *History, Man, and Reason: A Study in 19th-Century Thought*. Baltimore: Johns Hopkins University Press.

Marcuse, H. (1964). *One Dimensional Man*. Boston: Beacon Press.

Margalit, A. (1986). "The Past of an Illusion," in E. Ullman-Margalit (ed.), *The Kaleidoscope of Science*. Dordrecht: D. Reidel, pp. 89–94.

Marks, J. (1983). *Science and the Making of the Modern World*. London: Heinemann.

Mendelsohn, E. (1964). "Explanation in 19th Century Biology," in R. Cohen and M. Wartofsky (eds.), *Boston Studies in the Philosophy of Science*, vol. 2. Dordrecht: D. Reidel, pp.127–50.

Mendelsohn, E. (1974). "Revolution and Reduction: The Sociology of Methodological and Philosophical Concerns in 19th-Century Biology" in Y. Elkana (ed.), *Interaction Between Science and Philosophy*. New York: Humanities Press, pp. 407–27.

Mendelsohn, E. (1989). "Robert K. Merton: The Celebration and Defence of Science," *Science in Context* 3: 269–99.

Merton, R. K. (1977). *The Sociology of Science*. Chicago: University of Chicago Press.

Merz, J. T. (1965). *A History of European Thought in the 19th Century*, 4 vols (orig. 1896–1914). New York: Dover.

Mirowski, P. (2002). *Machine Dreams: Economics Becomes a Cyborg Science*. Cambridge, UK: Cambridge University Press.

Mirowski, P. (2004). *The Effortless Economy of Science?* Durham, NC: Duke University Press.

Mooney, C. (2005). *The Republican War on Science*. New York: Simon & Schuster.

More, M. (2005). "The Proactionary Principle." <http://www.maxmore.com/proactionary.htm>.

Newell, W. H. (ed.) (1998). *Interdisciplinarity: Essays from the Literature*. New York: The College Board.

Noble, D. (1997). *The Religion of Technology: The Divinity of Man and the Spirit of Invention*. New York: Alfred Knopf.

Noelle-Neumann, E. (1981). *The Spiral of Silence*. Chicago: University of Chicago Press.

Nowotny, H. et al. (2000). *Re-Thinking Science*. Cambridge, UK: Polity.

Office of Technology Assessment [OTA] (1991). *Federally Funded Research: Decisions for a Decade*. Washington, DC: US Government Printing Office.

O'Neill, J. (ed.) (1973). *Modes of Individualism and Collectivism*. London: Heinemann.

Pachter, H. (1984). *Socialism in History*. New York: Columbia University Press.

Page, S. (2006). "Path Dependence," *Quarterly Journal of Political Science* 1: 87–115.

Passmore, J. (1966). *A Hundred Years of Philosophy*, 2nd edn (orig. 1957). London: Duckworth.

Passmore, J. (1970). *The Perfectibility of Man*. London: Duckworth.

Pettit, P. (1997). *Republicanism: A Theory of Freedom and Government*. Oxford: Oxford University Press.

Piaget, J. (1976). *Psychology and Epistemology*. Harmondsworth, UK: Penguin.

Pickering, A. (1995). *The Mangle of Practice*. Chicago: University of Chicago Press.

Pinker. S. (2002). *The Blank Slate: The Modern Denial of Human Nature*. New York: Vintage.

Polanyi, K. (1944). *The Great Transformation*. Boston: Beacon Press.

Polanyi, M. (1957). *Personal Knowledge*. Chicago: University of Chicago Press.

Popper, K. (1945). *The Open Society and Its Enemies*. London: Routledge & Kegan Paul.

Popper, K. (1957). *The Poverty of Historicism*. New York: Harper & Row.

Popper, K. (1959). *The Logic of Scientific Discovery* (orig. 1935). New York: Harper & Row.

Popper, K. (1972). *Objective Knowledge*. Oxford: Oxford University Press.

Proctor, R. (1988). *Racial Hygiene: Medicine under the Nazis*. Cambridge, MA: Harvard University Press.

Proctor, R. (1991). *Value-Free Science? Purity and Power in Modern Knowledge*. Cambridge, MA: Harvard University Press.

Putnam, H. (1975). *Mind, Language and Reality*. Cambridge, UK: Cambridge University Press.

Pyenson. L. and Sheets-Pyenson, S. (1999). *Servants of Nature*. New York: Norton.

Pylyshyn, Z. (1984). *Computation and Cognition*. Cambridge, MA: MIT Press.

Quine, W. V. O. (1953). *From a Logical Point of View*. New York: Harper & Row.

Rabinbach, A. (1990). *The Human Motor: Energy, Fatigue, and the Origins of Modernity*. New York: Basic Books.

Radder, H. (2000). "Review of Steve Fuller, *The Governance of Science*," *Science, Technology & Human Values* 25: 520–7.

Ravetz, J. (1971). *Scientific Knowledge and Its Social Problems*. Oxford: Oxford University Press.

Reichenbach, H. (1938). *Experience and Prediction*. Chicago: University of Chicago Press.

Reisch, G. (2005). *How the Cold War Transformed the Philosophy of Science*. Cambridge. UK: Cambridge University Press.

Remedios, F. (2003). *Legitimizing Scientific Knowledge: An Introduction to Steve Fuller's Social Epistemology*. San Francisco: Lexington Books.

Rescher, N. (1978). *Peirce's Philosophy of Science*. South Bend: Notre Dame Press.

Restivo, S. (1983). "The Myth of the Kuhnian Revolution," in R. Collins, *Sociological Theory*. San Francisco: W. H. Freeman, pp. 293–305.

Richards, J. (ed.) (2002). *Are We Spiritual Machines?* Seattle: Discovery Institute.

Roco, M. and Bainbridge, W. S. (eds.) (2002). *Converging Technologies for Enhancing Human Performance: Nanotechnology, Biotechnology, Information Technology and Cognitive Science*. Arlington, VA: US National Science Foundation.

Rogers, E. (1962). *The Diffusion of Innovations*. New York: Free Press.

Roll-Hansen, N. (2005). *The Lysenko Effect: The Politics of Science*. Amherst, NY: Prometheus Books.

Rosenberg, A. (1994). *Instrumental Biology or the Disunity of Science*. Chicago: University of Chicago Press.

Rosenberg, N. (1972). *Technology and American Economic Growth*. New York: Harper & Row.

Rostow, W. W. (1960). *The Stages of Economic Growth: A Non-Communist Manifesto*. Cambridge, UK: Cambridge University Press.

Rouse, J. (2002). *How Scientific Practices Matter: Reclaiming Philosophical Naturalism*. Chicago: University of Chicago Press.

Ruse, M. (1979). *The Darwinian Revolution: Science Red in Tooth and Claw*. Chicago: University of Chicago Press.

Ruse, M. (2005). "Evolutionary Biology and the Question of Trust," in Koertge (2005), pp. 99–119.

Sassower, R. (1997). *Technoscientific Angst: Ethics and Responsibility*. Minneapolis: University of Minnesota Press.

Sassower, R. (2006). *Popper's Legacy: Rethinking Politics, Economics and Society*. Chesham, UK: Acumen Publishing.

Schaffer, S. (1991). "The Eighteenth Brumaire of Bruno Latour," *Studies in History & Philosophy of Science* 22: 174–92.

Schnaedelbach, H. (1984). *Philosophy in Germany, 1831–1933*. Cambridge, UK: Cambridge University Press.

Schumpeter, J. (1950). *Capitalism, Socialism and Democracy*, 2nd edn (orig. 1942). New York: Harper and Row.

Schwartzman, P. (1995). "The Population Growth Debate in the Public Sphere," *Social Epistemology* 9: 289–310.

Segerstrale, U. (2000). *Defenders of the Truth*. Oxford: Oxford University Press.

Serres, M. (1982). *Hermes*. Baltimore: Johns Hopkins University Press.

Shahidullah, S. (1991). *Capacity-Building in Science and Technology in the Third World*. Boulder, CO: Westview Press.

Shapere, D. (1984). *Reason and the Search for Knowledge*. Dordrecht: Kluwer.

Shapin, S. (1994). *The Social History of Truth*. Chicago: University of Chicago Press.

Shapin, S. (1996). *The Scientific Revolution*. Chicago, IL: University of Chicago Press.

Shearmur, J. (1996). *The Political Thought of Karl Popper*. London: Routledge.

Simon, H. (1977). *The Sciences of the Artificial*. Cambridge. MA: MIT Press.

Singer, P. (1975). *Animal Liberation*. New York: Random House.

Singer, P. (1999). *A Darwinian Left: Politics, Evolution and Cooperation*. London: Weidenfeld & Nicolson.

Sismondo, S. (2004). *An Introduction to Science and Technology Studies*. Oxford: Blackwell.

Smith, J. M. (1998). *Shaping Life: Genes, Embryos and Evolution*. London: Weidenfeld & Nicolson.

Smocovitis, V. B. (1996). *Unifying Biology: The Evolutionary Synthesis and Evolutionary Biology*. Princeton: Princeton University Press.

Sokal, A. and Bricmont, J. (1998). *Intellectual Impostures*. London: Phaidon.

Spengler, O. (1991). *The Decline of the West* (abridged edn) (orig. 1918). Oxford: Oxford University Press.

Standage, T. (1998). *The Victorian Internet*. London: Phoenix Press.

Staudenmaier. J. (1985). *Technology's Storytellers*. Cambridge, MA: MIT Press.

Stehr, N. (1994) *Knowledge Societies*. London: Sage.

Strum, S. and Fedigan, L. (eds.) (2000). *Primate Encounters: Models of Science, Gender and Society*. Chicago: University of Chicago Press.

Voegelin, E. (1968). *Science, Politics, and Gnosticism*. Chicago: Regnery Publishing.

Waddington, C. (1944). *The Scientific Attitude*. Harmondsworth, UK: Penguin.

Wallerstein, I. (ed.) (1996). *Open the Social Sciences*. Palo Alto: Stanford University Press.

Webster, C. (1975). *The Great Instauration: Science, Medicine and Reform, 1620–1660*. London: Duckworth.

Weikart, R. (2005). *From Darwin to Hitler*. New York: Macmillan.

Weinberg, S. (1992). *Dreams of a Final Theory*. New York: Pantheon.

Weinberg, S. (2001). *Facing Up: Science and Its Cultural Adversaries*. Cambridge, MA: Harvard University Press.

Werskey, G. (1988). *The Visible College: Scientists and Socialists in the 1930s*, 2nd edn (orig. 1978). London: Free Association Books.

White, H. (1972). *Metahistory*. Baltimore: Johns Hopkins University Press.

Whiten, A. and Byrne, R. (eds.) (1997). *Machiavellian Intelligence II: Extensions and Evaluations*. Cambridge, UK: Cambridge University Press.

Whiteside, K. (2002). *Divided Natures: French Contributions to Political Ecology*. Cambridge, MA: MIT Press.

Wiener, N. (1948). *Cybernetics: or Control and Communication in the Animal and the Machine*. Cambridge, MA: MIT Press.

Wilson, E. O. (1975). *Sociobiology: The New Synthesis*. Cambridge, MA: Harvard University Press.

Wilson, E. O. (1984). *Biophilia*. Cambridge, MA: Harvard University Press.
Wilson, E. O. (1992). *The Diversity of Life*. Cambridge, MA: Harvard University Press.
Wilson, E .O. (1998). *Consilience: The Unity of Knowledge*. New York: Alfred Knopf.
Wilson, E. O. (2002). *The Future of Life*. New York: Alfred Knopf.
Winner, L. (1977). *Autonomous Technology*. Cambridge, MA: MIT Press.
Wolfram, S. (2002). *A New Kind of Science*. Urbana, IL: Wolfram Media.
Woolgar, S. (ed.) (1988). *Knowledge and Reflexivity*. Sage: London.
Wuthnow, R. (1989). *Communities of Discourse*. Cambridge, MA: Harvard University Press.
Zagzebski, L. and Fairweather, A. (eds.) (2001). *Virtue Epistemology: Essays on Epistemic Virtue and Responsibility*. Oxford: Oxford University Press.

Index